Facing up to Nuclear Power

A Contribution to the Debate
on the Risks and Potentialities
of the Large-Scale Use of Nuclear Energy

Edited by
JOHN FRANCIS AND PAUL ABRECHT

THE SAINT ANDREW PRESS : EDINBURGH

First published in 1976
in Great Britain by
THE SAINT ANDREW PRESS
121 George Street, Edinburgh
and in the U.S.A. by
THE WESTMINSTER PRESS
Witherspoon Building, Philadelphia

ISBN 0 7152 0340 1

Printed and bound in Great Britain by
Morrison & Gibb Ltd, Edinburgh and London

FACING UP TO NUCLEAR POWER

CONTENTS

Section 4

SOCIAL ETHICS OF NUCLEAR POWER

Section 5

ECUMENICAL HEARING ON NUCLEAR
ENERGY: A Report to the Churches

Section 6

A STEP FORWARD

LIST OF CONTRIBUTORS

Dr. Paul Abrecht, Studies on Church and Society, World Council of Churches, Geneva

Prof. Hannes Alfvén, The Royal Institute of Technology, Department of Plasma Physics, Stockholm

Prof. H. B. G. Casimir, President of the European Physical Society

Dr. John M. Francis, Senior Research Fellow in Energy Studies, Heriot-Watt University, Edinburgh

Prof. Wolf Häfele, Deputy Director, IIASA, Austria

Dr. Shem Arungu-Olende, U.N. Economic Commission for Africa

Dr. Jan Prawitz, Special Assistant for Disarmament to the Minister of Defence, Sweden

Dr. Jorge Sabato, former Member of the Atomic Energy Commission, Argentina

Prof. Roger L. Shinn, Professor of Social Ethics, Union Theological Seminary, New York

Prof. Gérard Siegwalt, Professor of Dogmatics, University of Strasbourg

Dr. K. T. Thomas, Director of the Engineering Services Group, Bhabha Atomic Research Centre, Trombay, India

Dr. Alvin M. Weinberg, Oak Ridge National Laboratory, Tennessee

PUBLISHER'S NOTE

The majority of the text of *Facing up to Nuclear Power* has been photographed directly from *Anticipation*, No. 20 (May 1975) and No. 21 (October 1975), published by The World Council of Churches, Geneva.

FOREWORD

"Every new invention expands space, shortens time and destroys community."

(Attributed to Rosenstock-Huessy)

Nuclear energy, more than any previous invention of a scientific and technological culture, epitomizes the dilemma of infinite potential benefit coupled with infinite risk to the community at large. In a historical sense the introduction of new science and technology has seemed to offer more tangible benefits than costs in industrialized societies. In present circumstances the costs and benefits of nuclear energy now seem more ambiguous than earlier technical developments, particularly when seen in the context of the developing world. Is there something special about this branch of advanced technology which distinguishes it from all previous science-based technology? Is the balance point between costs and benefits sustainable into the long-term future? What is our Christian evaluation of the ethical and theological problems it poses?

In June 1974 at the World Council of Churches' Conference on *Science and Technology for Human Development* (Bucharest, Romania), the issue was posed as follows:

"It remains an open question whether the widespread proliferation of nuclear power plants is a desirable choice for society to make. Yet throughout the Western world, there are already clear signs of a growing dependence on this form of energy production. Since national energy policies are prepared in isolation, it is doubtful that the collective

international impact of these rapidly expanding construction programmes has been systematically assessed... A realistic appraisal of the implications of the nuclear process would seem to indicate unprecedented demands on capital, mineral resources and industrial capabilities, increasingly centralized systems of production, and a potentially large-scale environmental impact. The nuclear option entails a risk whose magnitude is a matter of debate. Widely divergent views of this risk are held in the scientific community."

In response, the Central Committee of the World Council, meeting in West Berlin, August 1974, requested the Sub-Unit on Church and Society to make an assessment of "the risks and potentialities of the expansion of nuclear power". After consultation with many persons active in the nuclear debate it was decided to convene a Hearing on Nuclear Energy bringing together nuclear scientists, scientists from related disciplines, technologists and politicians, as well as theologians and church leaders. The Hearing was held at Sigtuna, Sweden, June 24–29, 1975, at the invitation of the Swedish churches who helped to make the meeting possible.

The participants in the Hearing Group reflected the wide range of views on energy problems in general and on nuclear energy in particular. Their diverging positions were outlined in a preparatory prospectus for the Hearing which provided the themes for the meeting. In addition to background papers circulated in advance, a number of the participants were invited to submit papers on specific aspects of the issue which were discussed in plenary. The Hearing also benefited from the advice of a number of resource persons, representing various international agencies.

This book provides a balanced reconstruction of the work completed by the Hearing Group at Sigtuna. The initial prospectus for the meeting has been included together with a selection of the background papers. Alvin Weinberg's article on

"Social Institutions and Nuclear Energy," first published in *Science*, July 7, 1972, started a discussion of the "Faustian bargain" between the nuclear scientists and society. Dr. Weinberg is probably one of the leading advocates of nuclear power systems, having for many years been Director of the Oak Ridge National Laboratory in Tennessee, U.S.A. His article, only slightly shortened here, is the text of his Rutherford Centennial Lecture, presented at the annual meeting of the American Association for the Advancement of Science, in Philadelphia, December 1971.

The second essay, entitled "Hypotheticality and the New Challenges: The Pathfinder Role of Nuclear Energy," is an original interpretation of the nuclear safety issue by Professor Wolf Häfele. This essay first appeared in the July 1974 number of *Minerva*, a science quarterly published in the U.K., and is published here with the permission of the editor and the author. Professor Häfele is currently directing a major project on "Energy Systems" at the International Institute for Applied Systems Analysis, Laxenburg, Austria.

In addition to this perspective from two leading nuclear scientists, several important papers are included giving the views of Third World energy specialists.

The essays in this section on "The Nuclear Option" are by Dr K. T. Thomas, who is Director of the Engineering Services Group, Bhabha Atomic Research Centre, Trombay, India, and Dr Jorge Sabato, a former member of the Atomic Energy Commission in Argentina.

As a direct result of large increases in the price of crude oil in the world petroleum market, many countries now stand on the threshold of nuclear power development programmes. At an international level the monitoring and safeguards that must be applied in order to secure the nuclear fuel cycle will need to be kept under constant review. At Sigtuna, the relationship between the civil uses of nuclear energy and the control of nuclear weapons was identified in a concise statement presented to the Hearing Group by Jan Prawitz, special assistant for disarmament to the Minister of Defence in Sweden.

The prospects for alternative energy sources are dealt with in a summary form in the paper by Hannes Alfvén, a 1970 Nobel Laureate in Physics from the Royal Institute of Technology, Stockholm. This is largely based on his presentation to the 23rd Pugwash Conference on Science and World Affairs at Aulanko, Finland, in the autumn of 1973. This paper is followed by an African perspective of the wider energy options by Dr Shem Arungu-Olende, who is currently at the Centre for Natural Resources, Energy and Transport, United Nations, New York.

The ethical and theological significance of the debate on nuclear energy is presented in papers by Roger Shinn and Gérard Siegwalt. Both of these writers suggest a number of different ways in which the dominant technological culture, as reflected in the widespread adoption of nuclear power systems, may be required to give ground in the face of an informed public discussion.

The closing sections of the book incorporate the full text of the Ecumenical Hearing on Nuclear Energy, together with some responses that have been collated during the period that it has been in circulation through the direct agency of the World Council of Churches.

The book has been put together in such a form that it cannot be taken to be the last word. In truth, the real debate over the future role of nuclear energy has scarcely begun. The report of the Hearing Group does not appear as a consensus statement. What has resulted is the first serious attempt to bring the many sides of the argument together around one table and to ensure that each specific aspect of public concern has been freely and fairly identified.

Meantime, the debate continues.

SECTION 1

PLAN FOR AN ECUMENICAL HEARING ON THE RISKS AND POTENTIALITIES OF THE FURTHER EXPANSION OF NUCLEAR POWER PROGRAMMES

I. ORIGIN AND PURPOSE OF THE HEARING

A number of factors have converged to increase the urgency of broader public examination of the consequences of the expansion of nuclear power. The oil embargo of 1973, the energy crisis linked with the rapid rise in the price of oil, and the possible exhaustion of fossil fuels have obliged many nations to press the search for alternative sources of energy, leading governments to move towards a more rapid development of nuclear power programmes than envisaged earlier.

Present estimates by the IAEA (International Atomic Energy Agency) and other bodies speak of accelerated increases of nuclear power production by 1985, and beyond that by the turn of the century. This would mean a large increase in the number of nuclear energy plants by 1985, a prospect which has aroused the concern of many scientists and lay persons alike.

As Dr Sigvard Eklund, Director General of the IAEA noted in his statement to the 18th Regular Session of the General Conference of the Agency (October 16, 1974): "In some countries there is, to put it mildly, a deep reluctance to accept nuclear power as a way out of the present difficulties."

During preliminary contacts and conversations it was clear that a proper evaluation of the factors which will influence public acceptibility of expanded nuclear power cannot be easily or quickly accomplished. Moreover a number of organizations, including the IAEA and other U.N. agencies are already deploying considerable efforts to assess the risk and provide the proper safeguards. There is no doubt that the ethical values and social guidelines implicit in the present debate need to be clarified and that the churches might have a constructive contribution to make at this point. Church members, like almost everyone else, are only beginning to understand the enormous implications of the future of nuclear power development on a world-wide basis. This assessment by the World Council of Churches will help to encourage substantial discussion in the churches during the follow-up to the Nairobi Assembly. It may also contribute to the discussion of nuclear policies for the future which will be the concern of the next major IAEA Conference on Nuclear Power and its Fuel Cycle in 1977.

II. FIVE DIVERGENT VIEWS

Such a preliminary assessment by the churches will not involve a major research study but an evaluation and synthesis of the divergent views that can now be identified in many countries. A brief description of essential elements of five positions emerging from the present debate is given here. It is not suggested that there are simply five distinct positions; in short form, this is supposed to be a sketch map that gives some indication of the many shades of grey rather than a black-and-white distinction "for" or "against". Note that it has been necessary to condense some of the major arguments.

A. WHOLLY NEGATIVE

At this extreme position at one end of the spectrum of opposing views, it is possible to group together a number of ethical, political, economic and technical concerns. These can be described as follows:

1. Social/Ethical Dimension

It can be argued that future generations will inherit the problems of nuclear power development, particularly those relating to the safe management of obsolete reactor systems or the prolonged storage of high specific activity radioactive wastes. To the extent that unborn children will have to bear a significant fraction of the costs of energy production for the present era, there is moral obligation on the operators to secure all loopholes in the process of designing and engineering these systems. This aspect must not be underplayed for economic reasons.

2. Political Concerns

In many countries, nuclear development programmes will remain under direct Government control, with the result that the planning of large nuclear programmes may not be subject to independent appraisal or any degree of outside comment. The defence status of some of these installations will inevitably bring down a security blanket once the reactors are operational. The institutionalized separation of these plants may be necessary even in democratic states, but this does not necessarily lead to confidence in the overall political control of the technology.

3. Nuclear Safeguards at an International Level

There is a body of opinion which believes that the present scale of international machinery for administering nuclear safeguards and monitoring national programmes is already inadequate in the light of prospective sales of nuclear power reactors. The need for some immediate improvement in the situa-

8

tion through regional centres of control rather than a single centralized authority is now apparent.

4. Economics of Nuclear Power Generation

A new school of energy analysis has advanced the view that while the economic viability of nuclear power in competition with other systems of electricity generation is now proven, the overall energy costs in a rapidly expanding nuclear programme can offset some of the obvious advantages. This can best be illustrated in the following diagrams (Chapman, 1974):[1]

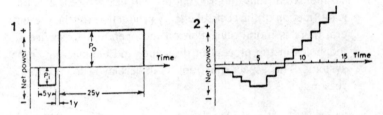

Figure 1: Energy consumption during building then energy production afterwards for a nuclear reactor with an energy ratio (E_R) of 10.

Figure 2: Variation in net power when a reactor of $E_R = 10$ is ordered each year.

According to this approach, a linear increase in the growth of nuclear power (one reactor ordered per year) indicates that it takes 13 years to produce a net energy output. There is an equally valid argument which suggests that the decommissioning costs of large nuclear installations might, at present rates of inflation, correspond to as much as 80% of the initial capital costs of construction.

5. Economic Self-Reliance must be based on Equilibrium Sources of Energy

In the developing world, it is very debatable whether the introduction of nuclear power can be defended in terms other than the pragmatic response of politicians threatened by an increasing balance-of-payments problem on the oil imports account. The smaller decentralized systems of power generation based on solar energy, wind, geothermal and hydroelectric power are more compatible with the manpower training programmes in most of these countries and should not be dismissed since they represent the best long-term prospect of energy self-sufficiency at the community level. This viewpoint is in line with the "technological opposition" who contend that alternative sources of energy will be developed, thereby replacing the need for widely proliferating nuclear power programmes.

B. PARTIALLY NEGATIVE

This is a discriminatory view based upon the merits of different fuel cycles, and consequently it reflects a more detailed knowledge of the design variables ranging from the current evaluation of thermal reactor types to the more advanced fast breeder concepts.

1. Detailed Cost Analysis

It is accepted that long-term factors, such as the costs of waste management (10 half lives of ^{239}Pu = 10 x 24,000 yrs.), should be properly reflected in the total costs of fuel recycling. New technological solutions to this kind of problem should also be examined in terms of the energy cost element and included in the overall energy balance of the particular nuclear power system.

2. Environmental Impact

The assessment of the impact of large energy projects on the natural environment is a comparatively recent requirement inside the planning legislation of even the industrialized countries. In a strictly operational sense (day-to-day), large nuclear power stations may compare quite favourably with conventional power stations utilizing fossil fuels, because of absence of large direct emissions of combustion products. Thermal pollution is acknowledged to be a considerable restraint on the siting of all types of centralized power stations. Under accident conditions (such as a loss-of-coolant accident), nuclear power systems introduce an additional dimension of risk that should be more openly assessed and generally understood by the public at large. While there may be specialist design solutions, the operational reliability of such systems must also be validated.

3. Security Risk

Responsibility for the security of nuclear operations resides with the licensing authority in each country. Under normal circumstances, this arrangement will probably function adequately, but in the present world, the threat of external terrorist activities cannot be lightly discounted. Since it is the transport of nuclear fuel materials between reactors and fuel reprocessing facilities that represent the most vulnerable linkages in the chain of operations, special security arrangements alone may not be sufficient to contain the risk of diversion. A total review of the nuclear handling systems in each country is now a matter of first priority.

4. Enlarged Plutonium Hazard for Breeder Reactors

Many technical experts in the nuclear power field have substantial reservations concerning the widespread development of fast breeder reactors incorporating a plutonium fuel cycle. In several countries, prototype fast reactors are already operational, so that with increasing pressure on the world uranium market, it seems inevitable that the higher conversion factors available in a $^{238}U/^{239}Pu$ breeding cycle will prove to be economically attractive. Substantial plutonium inventories, far exceeding present levels, will result in a new order of radiological hazard. The large scale introduction of this technology may be imminent; it will certainly remain as one of the most controversial aspects of nuclear power development in the period beyond 1985.

5. Rate of Nuclear Power Development

With countries in the industrialized and in the developing world already subject to considerable restraints in capital expenditure resulting from the present disturbances in the world monetary system, it can be reliably argued that energy conservation has become a matter of immediate priority in most cases. It is not clear to what extent this movement will constrain the growth of nuclear power programmes in view of the long lead time for construction that is now an accepted part of the forecasting and planning dilemma.

C. MIDDLE VIEW

The balance point between the economic attractions of low cost nuclear power from a range of available systems and the seemingly inevitable risks of increasing dependency on this highly complex technology will be central to the debate on future energy supplies. The feasibility of nuclear power in all its forms has been a preoccupation of industrialized countries for the past 30 years. Massive investment and commitment at the national level will not be simply passed over in response to pressure group activities from outside the industry. It would be wrong in many cases to deny that nuclear power has a major contribution to make in providing electricity for large conurbations. It would be equally wrong to imply a total dependency on nuclear power as the ultimate source of energy for the future. The middle view should advance a reasoned perspective of the place of nuclear power in the overall energy balance.

1. Synthesis

Although there are many variables and unknowns in the energy equation, a first approximation to the most reasonable balance between all available energy sources, including nuclear power, should be introduced into the debate. This will provide a general guideline on the projected significance of nuclear power development in contrast to the alternative sources, including fossil fuels, solar and geothermal energy.

2. Risk-Benefit Criteria

The introduction of social and environmental factors governing nuclear power alongside the orthodox economic criteria, will bring the analysis closer to the domain of public understanding and acceptability. Multidisciplinary teams sponsored by international agencies are probably best equipped to advance a general analytical framework which can then be adapted to national and regional situations. International standards on the siting criteria, ancillary planning legislation and licensing procedures for nuclear plant should be subject to scrutiny by an independent jury with no formalized commitment to nuclear power systems.

3. Nuclear Plant Operations Secured

Increased levels of security at the level of individual nuclear power plants will be complemented by shared responsibilities for processing fuels at international regional facilities. This will be essential if nuclear power plants are introduced in large numbers to countries in Latin America, Africa and Asia. Without access to localized reprocessing facilities, the transport problems would be formidable with a concomitant increase in the risk of diversion or accidental dispersion. This is a planning problem of a unique kind; within the boundaries of human fallibility, the engineering component can no longer be based on trial-and-error principles.

4. Management Strategies

a. A slowdown in the introduction of "advanced" nuclear systems, e.g., the liquid metal fast breeder reactor, will allow the consolidation of present day engineering competence in the nuclear power field. This strategy is particularly important if countries are to be persuaded **not** to jump the technology "gap" into this more uncertain end of the nuclear power spectrum.

b. While each country will seek to achieve the optional fuel mix that is most immediately compatible with its prospects for economic development, this does not necessarily lead to a zero energy growth situation in every case. There is nevertheless a wide margin of wasteful energy use in most industrialized countries that can be eliminated; the time-scale for achieving this is not likely to be less than 10-15 years.

c. Adaptation to new life styles requiring a different pattern of energy use will almost certainly become one of the principal challenges to people in all countries. In the process, it should be possible to introduce the alternative systems of energy production that are more in line with individual or community needs in many countries, particularly in the developing world. It remains to be seen whether the constraints may warrant a form of energy rationing that reflect this basic shift in the demand-supply equation.

D. PARTIALLY POSITIVE

1. Future Dependency of Developing Countries on Nuclear Power Development

As a result of the upward movement of the price of crude oil in world markets, the smaller scale of nuclear power plant (100-400 MW(e)) is now economically competitive with the conventional forms of electrical generation. In countries where there is already a declared commitment to nuclear power, this probably means an increased dependency over that already planned for the next decade; countries such as India and Brazil would appear to be in that category. There is also the possibility that some countries with a surplus of revenues from the petroleum markets may now be preparing to enter the nuclear power field.

2. Standardized Approach to Nuclear Power Development

There is a good deal of engineering wisdom behind the view that only a very standardized pattern of nuclear reactor types should be offered to countries on the threshold of a commitment to nuclear power. This would serve to minimize the uncertainty arising out of reactor performance in widely varying load-following situations. It would also enable the operators to become more familiar with safe-operating procedures and fuel handling requirements for a reactor series constructed over a fairly long period.

3. Investment and Manpower Training

It is obvious that both of these factors are highly interconnected. The capital demands for the successful implementation of a nuclear power programme are likely to be large and immediate. At the same time, it would be completely unworkable to embark upon the first phase of development without a suitable national catchment of scientific and engineering expertise in the major nuclear disciplines. This investment must equally reflect a capability for handling each stage in the fuel cycle at some reasonable date in the future, otherwise there is an implied dependency on the industrialized countries that are currently marketing the systems. In order to participate in the full economic benefits, investment in shared regional facilities for fuel reprocessing may be necessary between groups of countries at comparable stages of development.

E. WHOLLY POSITIVE

1. Nuclear Salesmanship

This is undoubtedly the province of the national nuclear corporation or the multi-national combine with an urgent need to recoup some of the massive research and development expenditures incurred in bringing a reactor concept to the market place. The prospect of an expanding market has failed to materialize time and time again during the past decade or so. While design safeguards and operational risks vary from one design to another, this is certainly not an area where aggressive salesmanship should be allowed a free interplay at the expense of the highest possible standards of international regulation and control.

2. No Reliable Alternatives

It will be strongly argued from this position that virtually all the alternative systems of energy production suggested for industrialized countries are far from proven in economic as well as in hard practical terms. To proceed without the benefits of widespread utilization of nuclear power would be considered both imprudent and totally unrealistic, given the scale of energy demand that the world is facing. Despite the finite element of risk, the place of nuclear power in most advanced economies is now guaranteed for that reason. The position is obviously not so clear cut in the developing world, but here again, nuclear power represents a hedge against uncertainty, if only because the viability of alternative energy sources has not been adequately demonstrated.

3. The Future Potential

If the standardized nuclear reactor package for developing countries is accepted as a workable concept, subject to the necessary implementation of international nuclear safeguards under the terms of the Non-Proliferation Treaty, then the world is embarked upon a broadly-based scenario for nuclear power development. There is a need to state that this path will conceivably lead on from thermal reactors to the fast breeder, from fission to fusion systems. Comparable hazards exist over the whole range of reactor types, but the trade-off could be the energy that humanity needs to survive in the quantities that reasonable men and women agree exists in theory.

III. ASSESSMENT OF DIVERGENT VIEWS

In assessing the merits of these views it was necessary in the course of the Hearing to make a systematic review and critique of six critical topics. Specialists were invited to prepare position papers on:

1. **Energy Planning:**
   Estimates of world energy requirements and consequences of possible shortages. Energy ratios and the search for a synthesis of energy sources.
   Economic Criteria: Evaluation of oil price movements and case for nuclear energy on economic and technological grounds; economic advantages of different energy systems.

2. **Energy Options for Developing Countries:**
   Presentation of African, Asian and Latin American energy needs.

3. **Risk Assessment:**
   Public acceptability criteria – based on comparison with other risks of a transient kind. The introduction of "permanent" damage to the biosphere as a new order of risk.

4. **Radiological Hazards:**
   Background radiation associated with large-scale nuclear power programmes; maximum permissible levels of radiation to the public at large. The assumptions concerning the siting of nuclear power complexes, fuel processing plants and long-term storage facilities in terms of the localized hazards.

5. **Operational Hazards:**
   Thermal pollution, waste management (especially the international aspects); the problems of maintaining a proper plutonium inventory with the increasing scale of nuclear power development. Standards for transportation of irradiated fuel ele-

ments. Plutonium diversion, the proliferation of nuclear materials, fuel reprocessing and the implementation of safeguards.

6. Ethical and Social Issues:

  a. Theological and ethical implications of "hypotheticality" or the "Faustian Bargain" as an attribute of nuclear technology. The morality of risk taking.
  b. The ethics of taking responsibility for others: Will nuclear power increase or diminish world social injustice by concentration of power? What is the theological basis of our concern for future generations?
  c. The conditions of survival and the quality of life: are there limits to technological ventures?

IV. SOCIAL AND ETHICAL IMPLICATIONS OF THE LONG-TERM COMMITMENT TO NUCLEAR POWER – AND THE CONTRIBUTION OF THE CHURCHES

In terms of the stages of nuclear development we are at the point of take-off. Unfortunately we do not understand sufficiently the chain of events arising out of these circumstances to be able to identify the system of international control and management that may be demanded of us in the future. It is also acknowledged that the future nuclear capacity will be beyond the scope of any group of governments or international agency, and that we have already embarked on this path without any clear means of controlling it, or of weighing its ultimate consequences for humanity at large.

The possibility exists therefore that the scale of operation and the proliferation of nuclear capacity would have consequences, because of the problem of security and safeguard, totally threatening the fabric of a democratic society as it is usually conceived. One of the proponents of nuclear power, Dr Alvin Weinberg, has openly acknowledged the vulnerability of such a large-scale dependency on nuclear technology:

> "We nuclear people have made a Faustian bargain with society. On the one hand we offer... an inexhaustible source of energy. Even in the short range, when we use ordinary reactors, we offer energy that is cheaper than energy from fossil fuel. Moreover, this source of energy, when properly handled, is almost non-polluting... But the price that we demand of society for this magical energy source is both a vigilance and a longevity of our social institutions that we are quite unaccustomed to."

This problem is already evident from the secretive and defensive attitudes of the operators of nuclear power installations when publicly called to account for the integrity of their operations. These attitudes are justified because at the present moment nuclear installations are recognized to be a risk.

We are going down a path where we have no understanding of the ultimate dangers. We have been overtaken by events (e.g. nuclear weapons and political agreements to supply nuclear plants around the world) so that the areas of choice may already be limited by purely strategic considerations. It is of course within the civil sector that public opinion has most recently been activated but this may be a suitable occasion to re-open the question of the ultimate implications.

We seem to have two choices:

1. To enter the "era of hypotheticality" (W. Häfele) enjoying the benefits of increased energy supplies, and accepting the uncertainties of the present day; but at the same time imposing a largely unknown risk on our descendants;

2. To regulate the introduction of this technology on a world-scale so that it is more in line with a clearly formulated understanding of the right measure of future dependency on nuclear energy – which today is still a largely unanswered question.

The fragility of the essential system of international economic and social interdependence has widely been demonstrated in recent years; to add a further dimension of risk and uncertainty and on such a large scale would seem to compound the problem. For this reason alone we must have a further understanding of the availability and use of nuclear power systems long before the situation becomes entirely irretrievable.

There is a need to approach these questions in a truly ecumenical spirit, that is, in the concern for the welfare of all peoples. The churches can contribute to the careful examination of the long-run implications of nuclear development. Some local groups of Christians are already deeply involved in local and national study and action on the further extension of nuclear power programmes. The Hearing sought to learn from their experience and contribute such guidance and information as may be available from the discussion of these questions at the international level.

The full text of the report from the Hearing Group is incorporated in Section 5.

V. THE WORK OF THE HEARING GROUP

The Hearing Group which was to meet together for the first time at Sigtuna in June 1975 had therefore been set a number of important tasks. From within its own constituency, it represented many elements in the broad spectrum of divergent views that were identified in the outline plan for the meeting. In the presence of expert witnesses who each gave a state-of-the-art report on the six major topics, the Group was then required to produce a collective judgement on the new circumstances governing nuclear power decisions in many countries throughout the industrialized and the developing world. Facing a seemingly impossible task to be completed within the course of six long hard days of argument and negotiation, the Group did succeed in producing a significant document as a record of the week's endeavour. The full text of the document is presented later in the book. It is in no sense a document striving for consensus within the Group on any specific topic. Such an agreement would serve no useful purpose and would only trivialize the profound differences of opinion which must exist wherever this vital subject is constructively debated. Nevertheless it was generally agreed that the final report does reflect fairly on the balance of technical and non-technical judgement exercised within the Group on the basis of the evidence compiled during the working sessions.

For the benefit of the general reader, some of the background papers and supporting material are included at this point in the book. This should allow the reader to clarify for himself or herself substantive arguments that could only be presented in outline in the main body of the report from the Hearing Group.

[1] P. Chapman, *New Scientist* 64 (no. 928), 19 December 1974.

SOCIAL INSTITUTIONS AND NUCLEAR ENERGY

ALVIN M. WEINBERG

Fifty-two years have passed since Ernest Rutherford observed the nuclear disintegration of nitrogen when it was bombarded with alpha particles. This was the beginning of modern nuclear physics. In its wake came speculation as to the possibility of releasing nuclear energy on a large scale: By 1921 Rutherford was saying, "The race may date its development from the day of the discovery of a method of utilizing atomic energy".[1]

Despite the advances in nuclear physics beginning with the discovery of the neutron by Chadwick in 1932 and Cockcroft and Walton's method for electrically accelerating charged particles, Rutherford later became a pessimist about nuclear energy. Addressing the British Association for the Advancement of Science in 1933, he said: "We cannot control atomic energy to an extent which would be of any value to do so".[2] Yet Rutherford did recognize the great significance of the neutron in this connection. In 1936, after Fermi's remarkable experiments with slow neutrons, Rutherford wrote: "... the recent discovery of the neutron and the proof of its extraordinary transmutations at very low velocities opens up new possibilities, if only a method could be found of producing slow neutrons in quantity with little expenditure of energy".[3]

Today the United States is committed to over 100×10^6 kilowatts of nuclear power, and the rest of the world to an equal amount. Rather plausible estimates suggest that by 2000 the United States may be generating electricity at a rate of 1000×10^6 kilowatts with nuclear reactors. Much more speculative estimates visualize an ultimate world of 15 billion people, living at something like the current U.S. standards: nuclear fission might then generate power at the rate of some 300×10^9 kilowatts of heat, which represents 1/400 of the flux of solar energy absorbed and reradiated by the earth.[4]

This large commitment to nuclear energy has forced many of us in the nuclear community to ask with the utmost seriousness questions which, when first raised, had a tone of unreality. When nuclear energy was small and experimental and unimportant, the intricate moral and institutional demands of a full commitment to it could be ignored or not taken seriously. Now that nuclear energy is on the verge of becoming our dominant form of energy, such questions as the adequacy of human institutions to deal with this marvellous new kind of fire must be asked, and answered, soberly and responsibly. In these remarks I review in broadest outline where the nuclear energy enterprise stands and what I think are its most troublesome problems; and I shall then speculate on some of the new and peculiar demands mankind's commitment to nuclear energy may impose on our human institutions.

Nuclear Burners – Catalytic and Noncatalytic

Even before Fermi's experiment at Stagg Field on December 2, 1942, reactor designing had captured the imagination of many physicists, chemists, and engineers at the Chicago Metallurgical Laboratory. Almost without exception, each of the two dozen main reactor types developed during the following 30 years had been discussed and argued over during those frenzied war years. Of these various reactor types, about five, moderated by light water, heavy water, or graphite, have survived. In addition, breeders, most notably the sodium-cooled plutonium breeder, are now under active development.

Today the dominant reactor type uses enriched uranium oxide fuel, and is moderated and cooled by water at pressures of 100 to 200 atmospheres. The water may generate steam directly in the reactor (so-called boiling water reactor (BWR)) or may transfer its heat to an external steam generator (pressurized water reactor (PWR)). These light water reactors (LWR) require enriched uranium and therefore at first could be built only in countries such as the United States and U.S.S.R., which had large plants for separating uranium isotopes.

In countries where enriched uranium was unavailable, or was much more expensive than in the United States, reactor development went along directions that utilized natural uranium: for example, reactors developed in the United Kingdom and France were based mostly on the use of graphite as moderator; those developed in Canada used D_2O as moderator. Both D_2O and graphite absorb fewer neutrons than does H_2O and therefore such reactors can be fuelled with natural uranium. However, as enriched uranium has become more generally available (of the uranium above ground, probably more by now has had its normal isotopic ratio altered than not), the importance of the natural ^{235}U isotopic abundance of 0.71% has faded. All reactor systems now tend to use at least slightly enriched uranium since its use gives the designer more leeway with respect to materials of construction and configuration of the reactor.

The PWR was developed originally for submarine propulsion where compactness and simplicity were the overriding considerations. As one who was closely involved in the very early thinking about the use of pressurized water for submarine propulsion (I still remember the spirited discussions we used to have in 1946 with Captain Rickover at Oak Ridge over the advantages of the pressurized water system), I am still a bit surprised at the enormous vogue of this reactor type for civilian power. Compact, and in a sense simple, these reactors were; but in the early days we hardly imagined that separated ^{235}U would ever be cheap enough to make such reactors really economical as sources of central station power.

Four developments proved us to be wrong. First, separated ^{235}U which at the time of Nautilus cost around $100 per gram fell to $12 per gram. Second, the price of coal rose from around $5 per ton to $8 per ton. Third, oxide fuel elements, which use slightly enriched fuel rather than the highly enriched fuel of the original LWR, were developed. This meant that the cost of fuel in a LWR could be, say, 1.9 mills per kilowatt hour (compared with around 3 mills per electric kilowatt hour for a coal-burning plant with coal at $8 per ton). Fourth, pressure vessels of a size that would have boggled our minds in 1946 were common by 1970: the pressure vessel for a large PWR may be as much as 8 1/2 inches thick and 44 feet tall. Development of these large pressure vessels made possible reactors of 1000 megawatts electric (MWe) or more, compared with 60MWe at the original Shippingport reactor. Since per unit of output a large power plant is cheaper than a small one, this increase in reactor size was largely responsible for the economic breakthrough of nuclear power.

Although the unit cost of water reactors has not fallen as much as optimists such as I had estimated, present costs are still low enough to make nuclear power competitive...

Water-moderated reactors burn ^{235}U, which is the only naturally occuring fissile isotope. But the full promise of nuclear fission will be achieved only with successful breeders. These are reactors that, essentially, burn the very abundant isotopes ^{238}U or ^{232}Th; in the process, fissile ^{239}Pu or ^{233}U acts as regenerating catalyst – that is, these isotopes are burned and regenerated. I therefore like to call reactors of this type *catalytic nuclear burners*. Since ^{238}U and ^{232}Th are immensely abundant (though in dilute form) in the granitic rocks, the basic fuel for such catalytic nuclear burners is, for all practical purposes, inexhaustible. Mankind will have a permanent source of supply once such catalytic nuclear burners are developed.

Most of the world's development of a breeder is centered around the sodium-cooled, ^{238}U burner in which ^{239}Pu is the catalyst and in which the energy of the neutrons is above 100×10^3 electron volts. No fewer than 12 reactors of this liquid metal fast breeder reactor (LMFBR) type are being worked on actively, and the United Kingdom plans to start a commercial prototype fast breeder by 1975. Some work continues on alternatives...

Nuclear Power and Environment

The great surge to nuclear power is easy to understand. In the short run, nuclear power is cheaper than coal power in most parts of the United States, in the long run, nuclear breeders assure us of an all but inexhaustible source of energy. Moreover, a properly operating nuclear power plant and its sub-systems (including transport, waste disposal, chemical plants, and even mining) are, except for the heat load, far less damaging to the environment than a coal-fired plant would be.

The most important emissions from a routinely operating reactor are heat and a trace of radioactivity. Heat emissions can be summarized quickly. The thermal efficiency of a PWR is 32%; that of a modern coal-fired power plant is around 40%. For the same electrical output the nuclear plant emits about 14% more waste heat than the coal plant does; in this one respect, present-day nuclear plants are more polluting than coal-fired plants. However, the higher temperature plants, such as the gas-cooled, the molten salt breeder, and the liquid metal fast breeder, operate at about the same efficiency as does a modern coal-fired plant. Thus, nuclear reactors of the future ought to emit no more heat than do other sources of thermal energy.

As for routine emission or radioactivity, even when the allowable maximum exposure to an individual at the plant boundary was set at 500 millirems (mrem) per year, the hazard, if any, was extremely small. But for practical purposes, technological advances have all but eliminated routine radioactive emission. These improvements are taken into account in the newly proposed regulations of the Atomic Energy Commission (AEC) requiring, in effect, that the dose imposed on any individual living near the plant boundary either by liquid or by gaseous effluents from LWR's should not exceed 5mrem per year. This is to be compared with the natural background which is around 100 to 200 mrem per year, depending on location, or the medical dose which now averages around 60 mrem per year.

As for the emissions from chemical reprocessing plants, data are relatively scant since but one commercial plant, the Nuclear Services Plant at West Valley, New York, has been operating, and this only since 1966. During this time, liquid discharges have imposed an average dose of 75 mrem per year at the boundary. Essentially no ^{131}I has been emitted. As for the other main gaseous effluents, all the ^{85}Kr and ^{3}H contained in the fuel has been released. This has amounted to an average dose from gaseous discharge of about 50 mrem per year.

Technology is now available for reducing liquid discharges, and processes for retaining ^{85}Kr and ^{3}H are being developed at AEC laboratories. There is every reason to expect these processes to be successful. Properly operating radiochemical plants in the future should emit no more radioactivity than do properly operating reactors – that is, less than 10% of the natural background at the plant boundary.

There are some who maintain that even 5 mrem per year represents an unreasonable hazard. Obviously there is no way to decide whether there is any hazard at this level. For example, if one assumes a linear dose-response for genetic effects, then to find, with 95% confidence, the predicted 0.5% increase in genetic effect in mice at a dose of, say, 150 mrem would require 8 billion animals. At this stage the argument passes from science into the realm of what I call trans-science, and one can only leave it at that.

My main point is that nuclear plants are indeed relatively innocuous, large-scale power generators if they and their subsystems work properly. The entire controversy that now surrounds the whole nuclear power enterprise therefore hangs on the answer to the question of whether nuclear systems can be made to work properly; or, if faults develop, whether the various safety systems can be relied upon to guarantee that no harm will befall the public.

The question has only one answer: there is no way to guarantee that a nuclear fire and all of its subsystems will never cause harm. But I shall try to show why I believe the measures that have been taken, and are being taken, have reduced to an acceptably low level the probability of damage.

I have already discussed low-level radiation and the thermal emissions from nuclear systems. Of the remaining possible causes of concern, I shall dwell on the three that I regard as most important: reactor safety, transport of radioactive materials, and permanent disposal of radioactive wastes.

Avoiding Large Reactor Accidents

One cannot say categorically that a catastrophic failure of a large PWR or a BWR and its containment is impossible. The most elaborate measures are taken to make the probability of such occurrence extremely small. One of the prime jobs of the nuclear community is to consider all events that could lead to accident, and by proper design to keep reducing their probability however small it may be. On the other hand, there is some danger that in mentioning the matter one's remarks may be misinterpreted as implying that the event is likely to occur.

Assessment of the safety of reactors depends upon two rather separate considerations: prevention of the initiating incident that would require emergency safety measures; and assurance that the emergency measures, such as the emergency core cooling, if ever called upon, would work as planned. In much of the discussion and controversy that has been generated over the safety of nuclear reactors, emphasis has been placed on what would happen if the emergency measures were called upon and failed to work. But to most of us in the reactor community, this is secondary to the question: How certain can we be that a drastic accident that calls into play the emergency systems will never happen? What one primarily is counting upon for the safety of a reactor is the integrity of the primary cooling system: that is, on the integrity of the pressure vessel and the pressure piping. Excruciating pains are taken to assure the integrity of these vessels and pipes. The watchword throughout the nuclear reactor industry is *quality assurance:* every piece of hardware in the primary system is examined, and reexamined, to guarantee insofar as possible that there are no flaws.

Nevertheless, we must deal with the remote contingency that might call the emergency system into action. How certain can one be that these will work as planned?

Three barriers prevent radioactivity from being released: fuel element cladding, primary pressure system, and containment shell. In addition to the regular safety system consisting primarily of the control and safety rods, there are elaborate provisions for preventing the residual radioactive heat from melting the fuel in the event of a loss of coolant. In the BWR there are sprays that spring into action within 30 seconds of an accident. In both the PWR and BWR, water is injected under pressure from gas-pressurized accumulators. In both reactors there are additional systems for circulating water after the system has come to low pressure, as well as means for reducing the pressure of steam in the containment vessel. This latter system also washes down or otherwise helps remove any fission products that may become airborne.

In analyzing the ultimate safety of a LWR, one tries to construct scenarios – improbable as they may be – of how a catastrophe might occur and then one tries to provide reliable countermeasures for each step in the chain of failures that could lead to catastrophe. The chain conceivable could go like this. First, a pipe might break, or the safety system might fail to respond when called upon in an emergency. Second, the emergency core cooling system might fail. Third, the fuel might melt, might react also with the water, and conceivably might melt through the containment. Fourth, the containment might fail catastrophically, if not from melt itself, then from missiles or overpressurization, and activity might then spread to be public. There may be other modes of catastrophic failure – for example, earthquakes or acts of violence – but the above is the more commonly identified sequence.

To give flavor of how the analysis of an accident is made, let me say a few words about the first and second steps of this chain. As a first step, one might imagine failure of the safety system to respond in an emergency, say, when the bubbles in a BWR collapse after a fairly routine turbine trip. Here the question is not that some safety rods will work and some will not, but rather that a common mode failure might render the entire safety system inoperable. Thus if all the electrical cables actuating the safety rods were damaged by fire, this would be a common mode failure. Such a common mode failure is generally regarded as impossible, since the actuating cables are carefully segregated, as are groups of safety rods, so as to avoid such an accident. But one cannot *prove* that a common mode failure is impossible. It is noteworthy that on September 30, 1970, the entire safety system of the Hanford-N reactor (a one-of-a-kind water-cooled, graphite-moderated reactor) did fail when called upon; however, the backup samarium balls dropped precisely as planned and shut off the reactor. One goes a long way toward making such a failure incredible if each big reactor, as in the case of the Hanford-N reactor, has two entirely independent safety systems that work on totally different principles. In the case of BWR, shutoff of the recirculation pumps in the all but incredible event the rods fail to drop constitutes an independent shutoff mechanism, and automatic pump shutoff is being incorporated in the design of modern BWR's.

The other step in the chain that I shall discuss is the failure of the emergency core cooling system. At the moment, there is some controversy whether the initial surge of emergency core cooling water would bypass the reactor or would in fact cool it. The issue was raised recently by experiments on a very small scale (9-inch-diameter pot) which indeed suggested that the water in that case would bypass the core during the blow-down phase of the accident. However, there is a fair body of experts within the reactor community who hold that these experiments were not sufficiently accurate simulations of an actual PWR to bear on the reliability or lack of reliability of the emergency core cooling in a large reactor.

Obviously the events following a catastrophic loss of coolant and injections of emergency coolant are complex. For example, one must ask whether the fuel rods will balloon and block coolant channels, whether significant chemical reactions will take place, or whether the fuel cladding will crumble and allow radioactive fuel pellets to fall out.

Such complex sequences are hardly susceptible to a complete analysis. We shall never be able to estimate everything that will happen in a loss-of-coolant accident with the same kind of certainty with which we can compute the Palmer series or even the course of the ammonia synthesis reaction in a fertilizer plant. The best that we can do as knowledgeable and concerned technologists is to present the evidence we have, and to expect policy to be based upon informed – not uninformed – opinion.

Faced with questions of this weight, which in a most basic sense are not fully susceptible to a yes or no scientific answer, the AEC has invoked the adjudicatory process. The issue of the reliability of the emergency core cooling system is being taken up in hearings before a special board drawn from the Atomic Safety and Licensing Board Panel. The record of the hearings is expected to contain all that is known about emergency core cooling systems and to provide the basis for setting the criteria for design of such systems.

Transport of Radioactive Materials

If, by the year 2000, we have 10^6 megawatts of nuclear power, of which two-thirds are liquid metal fast breeders, then there will be 7,000 to 12,000 annual shipments of spent fuel from reactors to chemical plants, with an average of 60 to 100 loaded casks in transit at all times. Projected shipments might contain 1.5 tons of core fuel which has decayed for as little as 30 days, in which case each shipment would generate 300 kilowatts of thermal power and 75 megacuries of radioactivity. By comparison, present casks from LWR's might produce 30 kilowatts and contain 7 megacuries.

Design of a completely reliable shipping cask for such a radioactive load is a formidable job. At Oak Ridge our engineers have designed a cask that looks very promising. As now conceived, the heat would be transferred to air by liquid metal or molten salt; and the cask would be provided with rugged shields which would resist deformation that might be caused by a train wreck. To be acceptable the shipping casks must be shown to withstand a 30-minute fire and a drop from 30 feet onto an unyielding surface.

Can we estimate the hazard associated with transport of these materials? The derailment rate in rail transport (in the United States) is 10^{-6} per car mile. Thus, if there were 12,000 shipments per year, each of a distance of 1000 miles, we would expect 12 derailments annually. However, the number of serious accidents would be perhaps 10^{-4} to 10^{-6}-fold less frequent; and shipping casks are designed to withstand all but the most serious accident (the train wreck near an oil refinery that goes into flames as a result of the crash). Thus the statistics – between 1.2×10^{-3} and 1.2×10^{-5} serious accidents per year – at least until the year 2000, look quite good. Nevertheless the shipping problem is a difficult one and may force a change in basic strategy. For example, we may decide to cool fuel from LMFBR's in place for 360 days before shipping: this reduces the heat load sixfold, and increases the cost of power by only around 0.2 mill per electric kilowatt hour. Or a solution that I personally prefer is to cluster fast breeders in nuclear power parks which have their own on-site reprocessing facilities.[5] Clustering reactors in this way would make both cooling and transmission of power difficult; also such parks would be more vulnerable to common mode failure, such as acts of war or earthquakes. These difficulties must be balanced against the advantage of not shipping spent fuel off-site, and of simplifying control of fissile material against diversion. To my mind, the advantages of clustering outweigh its disadvantages; but this again is a trans-scientific question which can only be adjudicated by a legal or political process, rather than by scientific exchange among peers.

By the year 2000, according to present projections, we shall have to sequester about 27,000 megacuries of radioactive wastes in the United States; these wastes will be generating 100,000 kilowatts of heat at that time...

The wastes will include about 400 megacuries of transuranic alpha emitters. Of these, the ^{239}Pu with a half-life of 24,000 years will be dangerous for perhaps 200,000 years.

Can we see a way of dealing with these unprecedentedly treacherous materials? I believe we can, but not without complication.

There are two basically different approaches to handling the wastes. The first, urged by W. Bennett Lewis of Chalk River,[6] argues that once man has opted for nuclear power he has committed himself to essentially perpetual surveillance of the apparatus of nuclear power, such as the reactors, the chemical plants, and others. Therefore, so the argument goes, there will be spots on the earth where radioactive operations will be continued in perpetuity. The wastes then would be stored at these spots, say in concrete vaults. Lewis further refines his ideas by suggesting that the wastes be recycled so as to limit their volume. As fission products decay, they are removed and thrown away as innocuous nonradioactive species; the transuranics are sent back to the reactors to be burned. The essence of the scheme is to keep the wastes under perpetual, active surveillance and even processing. This is deemed possible because the original commitment to nuclear energy is considered to be a commitment in perpetuity.

There is merit in these ideas; and indeed permanent storage in vaults is a valid proposal. However, if one wishes to perpetually rework the wastes as Lewis suggests, chemical separations would be required that are much sharper than those we now know how to do; otherwise at every stage in the recycling we would be creating additional low-level wastes. We probably can eventually develop such sharp separation methods; but these, at least with currently visualized techniques, would be very expensive. It is on this account that I like better the other approach which is to find some spot in the universe where the wastes can be placed forever out of contact with the biosphere. Now the only place where we know absolutely the wastes will never interact with man is in far outer space. But the roughly estimated cost of sending wastes into permanent orbit with foreseeable rocket technology is in the range of 0.2 to 2 mills per electric kilowatt hour, not to speak of the hazard of an abortive launch. For both these reasons I do not count on rocketing the wastes into space.

This pretty much leaves us with disposal in geologic strata. Of the many possibilities – deep rock caverns, deep wells, bedded salt – the latter has been chosen, at least on an experimental basis, by the United States and West Germany. The main advantages of bedded salt are primarily that, because salt dissolves in water, the existence of a stratum of bedded salt is evidence that the salt has not been in contact with circulating water during ecologic time. Moreover, salt flows plastically; if radioactive wastes are placed in the salt, eventually the salt ought to develop the wastes and sequester them completely.

These arguments were adduced by the National Academy of Sciences Committee on Radioactive Waste Management[7] in recommending that the United States investigate bedded salt (which underlies 500,000 square miles in our country) for permanent disposal of radioactive wastes. And, after 15 years of discussion and research, the AEC about a year ago decided to try large-scale waste disposal in an abandoned salt mine in Lyons, Kansas. If all goes as planned, the Kansas mine is to be used until A.D. 2000. What one does after A.D. 2000 would of course depend on our experience during the next 30 years (1970 to 2000). In any event, the mine is to be designed so as to allow the wastes to be retrieved during this time.

The salt mine is 1000 feet deep, and the salt beds are around 300 feet thick. The beds were laid down in Permian times and had been undisturbed, until man himself intruded, for 200 million years. Experiments in which radioactive fuel elements were placed in the salt have clarified details of the temperature distribution around the wastes, the effect of radiation on salt, the migration of water of crystallization within the salt, and so on.

The general plan is first to calcine the liquid wastes to a dry solid. The solid is then placed in metal cans, and the cans are buried in the floor of a gallery excavated in the salt mine. After the floor of the gallery is filled with cans, the gallery is backfilled with loose salt. Eventually this loose salt will consolidate under the pressure of the overburden, and the entire mine will be rescaled. The wastes will have been sequestered, it is hoped, forever.

Much discussion has centered around sequestration of just how certain we are that the events will happen exactly as we predict. For example, is it possible that the mine will cave in and that this will crack the very thick layers of shale lying between the mine and an aquifer at 500 feet below the surface? There is reason to suggest that this will not happen and I believe most, though not all geologists who have studied the matter agree that the 500-foot-thick layer of shale above the salt is too strong to crack so completely that water could invade the mine from above.

But man's interventions are not so easily disposed of. In Kansas there are some 130,000 oil wells and dry holes that have been drilled through these salt infusions. These holes penetrate aquifers and in principle they can let water into the mine. For the salt mine to be acceptible, one must plug all such holes. At the originally proposed site there were 30 such holes; in addition, solution mining was practised nearby. For this reason, the AEC recently authorized the Kansas State Geological Survey to study other sites that were not peppered with man-made holes. The AEC also announced recently its intention to store solidified wastes in concrete vaults, pending resolution of these questions concerning permanent disposal in geologic formations.

Man's intervention complicates the use of salt for waste disposal; yet by no means does this imply that we must give up the idea of using salt. In the first place, such holes can be plugged, though this is costly and requires development. In the second place, let us assume the all but incredible event that the mine is flooded – let us say 10,000 years hence. By that time, since no new waste will be placed in the mine after A.D. 2000, all the highly radioactive waste-decaying species, notably ^{90}Sr and ^{137}Cs would have decayed. The main radioactivity would then come from the alpha emitters. The mine would contain 38 tons of ^{239}Pu mixed with about a million tons of nonradioactive material. The plutonium in the cans is thus diluted to 38 parts per million; since plutonium is, per gram, 10,000 times more hazardous than natural uranium in equilibrium with the daughters, these diluted waste materials would present a hazard of the same order as an equal amount of pitchblende. Actually, the 38 tons of ^{239}Pu is spread over 200 acres. If all the salt associated with the ^{239}Pu were dissolved in water, as conceivably could result from total flooding of the mine, the concentration of plutonium in the resulting salt solution would be well below maximum permissible concentrations. In other words, by virtue of having spread the plutonium over an area of 200 acres, we have to a degree ameliorated the residual risk · in the most unlikely event that the mines are flooded.

Despite such assurances, the mines must not be allowed to flood, especially before the ^{137}CS and ^{90}Sr decay. We must prevent man from intruding – and this can be assured only by man himself. Thus we again come back to the great desirability, if not absolute necessity in this case, of keeping the wastes under some kind of surveillance in perpetuity. The great advantage of the salt method over, say, the perpetual reworking method, or even the aboveground concrete vaults without reworking, is that our commitment to surveillance in the case of salt is minimal. All we have to do is prevent man from intruding, rather than keeping a priesthood that forever reworks the wastes or guards the vaults. And if the civilization should falter, which would mean, among other things, that we abandon nuclear power altogether, we can be almost (but not totally) assured that no harm would befall our recidivist descendants of the distant future.

Social Institutions – Nuclear Energy

We nuclear people have made a Faustian bargain with society. On the one hand, we offer – in the catalytic nuclear burner – an inexhaustible source of energy. Even in the short range, when we use ordinary reactors, we offer energy that is cheaper than energy from fossil fuel. Moreover, this source of energy, when properly handled, is almost non-polluting. Whereas fossil fuel burners must emit oxides of carbon and nitrogen, and probably will always emit some sulfur dioxide, there is no intrinsic reason why nuclear systems must emit any pollutant – except heat and traces of radioactivity.

But the price that we demand of society for this magical energy source is both a vigilance and a longevity of our social institutions that we are quite unaccustomed to. In a way, all of this was anticipated during the old debates over nuclear power. As matters have turned out, nuclear weapons have stabilized at least the relations between the superpowers. The prospects of an allout Third World War seem to recede. In exchange for this atomic peace we have had to manage and control nuclear weapons. In a sense, we have established a military inadvertent use of nuclear weapons, which maintains what *a priori* seems to be a precarious balance between readiness to go to war and vigilance against human errors that would precipitate war. Moreover, this is not something that will go away, at least not soon. The discovery of the bomb has imposed an additional demand on our social institutions. It has called forth this military priesthood upon which in a way we all depend for our survival.

It seems to me (and in this I repeat some views expressed very well by Atomic Energy Commissioner Wilfrid Johnson) that peaceful nuclear energy probably will make demands of the same sort on our society, and possibly of even longer duration. To be sure, we shall steadily improve the technology of nuclear energy; but, short of developing a truly successful thermo-nuclear reactor, transport of radioactive materials, and waste disposal. And even if thermonuclear energy proves to be successful, we shall still have to handle a good deal of radioactivity.

We make two demands. The first, which I think is the easier to manage, is that we exercise in nuclear technology the very best techniques and that we use people of high expertise and purpose. Quality assurance is the phrase that permeates much of the nuclear community these days. It connotes using the highest standards of engineering design and execution; of maintaining proper discipline in the operation of nuclear plants in the face of the natural tendency to relax as a plant becomes older and more familiar; and perhaps of managing and operating our nuclear power plants with people of higher qualification than were necessary for managing and operating non-nuclear power plants; in short, of creating a continuing tradition of meticulous attention to detail.

The second demand is less clear, and I hope it may prove to be unnecessary. This is the demand for longevity in human institutions. We have relatively little problem dealing with wastes if we can assume always that there will be intelligent people around to cope with eventualities we have not thought of. If the nuclear parks that I mention are permanent features of our civilization, then we presumably have the social apparatus, and possibly the sites, for dealing with our wastes indefinitely. But even our salt mine may require some small measure of surveillance if only to prevent men in the future from drilling holes into the burial grounds.

Eugene Wigner has drawn an analogy between this commitment to a permanent social order that may be implied in nuclear energy and our commitment to a stable, year-in and year-out social order when man moved from hunting and gathering to agriculture. Before agriculture, social institutions hardly required the long-lived stability that we now take so much for granted. And the commitment imposed by agriculture in a sense was forever: the land had to be tilled and irrigated every year in perpetuity; the expertise required to accomplish this task could not be allowed to perish or man would perish; his numbers could not be sustained by hunting and gathering. In the same sense, though on a much more highly sophisticated plane, the knowledge and care that goes into the proper building and operation of nuclear power plants and their subsystems is something that we are committed to forever, so long as we find no other practical energy source of infinite extent.[8]

Let me close on a somewhat different note. The issue I have discussed here – reactor safety, waste disposal, transport of radioactive materials – are complex matters about which little can be said with absolute certainty. When we say that the probability of a serious reactor incident is perhaps 10^{-8} or even 10^{-4} per reactor per year, or that the failure of all safety rods simultaneously is incredible, we are speaking of matters that simply do not admit of the same order of scientific certainty as when we say it is incredible for heat to flow against a temperature gradient or for a perpetuum mobile to be built. As I have said earlier, these matters have transscientific elements. We claim to be responsible technologists, and as responsible technologists we give as our judgement that these probabilities are extremely – almost vanishingly – small; but we can never represent these things as certainties. The society must then make the choice, and this is a choice that we nuclear people cannot dictate. We can only participate in making it. Is mankind prepared to exert the eternal vigilance needed to ensure proper and safe operation of its nuclear energy system? This admittedly is a significant commitment that we ask of society. What we offer in return, an all but infinite source of relatively cheap and clean energy, seems to me to be well worth the price.

[1] "50 and 100 Years Ago", *Scientific American* 225, 10 (November 1971).

[2] J. Bartlett, *Familiar Quotations* (Little, Brown, Boston, ed. 14, 1968).

[3] E.N. da C. Andrade, *Rutherford and the Nature of the Atom* (Doubleday, Garden City, N.Y., 1964), p. 210.

[4] A.M. Weinberg and R.P. Hammond in *Proceedings of the Fourth International Conference on the Peaceful Uses of Atomic Energy* (United Nations, New York); *Bulletin of the Atomic Scientist* 28, 5, 43 (March 1972).

[5] A.M. Weinberg, "Demographic Policy and Power Plant Siting", Senate Interior and Insular Affairs Committee, Symposium on Energy Policy and National Goals, Washington, D.C., October 20, 1971.

[6] W.B. Lewis, *Radioactive Waste Management in the Long Term* (DM-123, Atomic Energy of Canada Ltd., Chalk River, July 13, 1971).

[7] *Disposal of Solid Radioactive Wastes in Bedded Salt Deposits* (National Academy of Sciences – National Research Council, Washington, D.C.,

1970; *Disposal of Radioactive Wastes on Land* (Publication 519, National Academy of Sciences – National Research Council, Washington, D.C., 1957); *Report to the U.S. Atomic Energy Commission* (National Academy of Sciences – National Research Council, Committee on Geologic Aspects of Radioactive Waste Disposal, Washington, D.C., May 1966).

[8] Prof. Friedrich Schmidt-Bleck of the University of Tennessee pointed out to me that the dikes of Holland require a similar institutional commitment in perpetuity.

HYPOTHETICALITY AND THE NEW CHALLENGES: THE PATHFINDER ROLE OF NUCLEAR ENERGY

WOLF HÄFELE

To those who for many years have been active in the promotion of nuclear energy, the opposition of the public to the large-scale application of peaceful nuclear energy has come as a surprise. The experience of public hearings and face-to-face discussions with the opponents of nuclear energy has made them aware of modes of thought and criteria of judgement which they had not encountered previously. It is now necessary to reflect on these alternative modes of thought and judgement in order to arrive at new ones, and, by so doing, to improve the basis for rational action.

Further, it appears that these alternative modes of thought and judgement and the responses which should be called forth by them do not arise on the occasion of large-scale uses of peaceful applications of nuclear energy alone. They are only adumbrations of a broader and more general development in thinking about science and technology. In this paper, I attempt to exemplify this development. I will begin with a brief account of the development of nuclear energy.

The Development of Nuclear Energy

The development of the peaceful applications of nuclear energy was related to its military applications in various degrees which changed with the times and which were different in various parts of the world. Originally basic scientific and technological research was to the forefront. But, more and more, the technological problems of large nuclear components, facilities and plants were the ones which attracted both scientists and engineers and funds for research and development. It was from this development of nuclear energy that the phenomenon of "big science" emerged as a new form of scientific organisation and as a new category of thought about science policy. The use of nuclear power for the production of electricity had to be competitive with fossil fuels, which in the 1950s and 1960s were temporarily cheap. It was thought that even if nuclear power turned out not to be fully competitive, fossil power would continue to be available inexpensively and in any amount desired. The drive for the development of nuclear power came also from an unquestioned belief in the desirability of technological innovation. This was particularly the case in the highly industrialised states without nuclear weapons.[1]

The order of the Oyster Creek nuclear power plant given to the General Electric Company by the Jersey Central Power and Light Company promoted the commercial breakthrough towards nuclear power. Since then, 150,000 MWe have been firmly ordered, have been under construction or in operation in the United States. The corresponding figures for Germany and Japan are 13,000 MWe and 15,000 MWe. Practically all of these are light water reactors (LWR's). In Great Britain and France, experience in the design, construction and operation of gas-cooled, graphite-moderated reactors, which demonstrates a remarkable technological capacity, has been considerable. However, it was not possible to translate this experience into a commercially feasible operation.

The development of reactors must be accompanied by the establishment of a fuel industry, the operation of which is commercially feasible. Since fuel elements have to be available prior to the operation of power plants, such a nuclear fuel industry is now being established. The problem here is the minimum size of the fabrication plant. To meet this requirement a single fabrication plant has to serve several thousand megawatts of nuclear power plant capacity. For many decades, enrichment will continue to be necessary since the present nuclear power plants make use of enriched uranium. Thus far, the gaseous diffusion process has been demonstrated as technologically and economically feasible. In the years to come, such feasibility may also be expected for the centrifuge process. In addition, the nozzle process must be mentioned. Again, a large capacity of nuclear power plants to be served is required to make enrichment a practical economic undertaking.

After irradiation in such nuclear power plants, the spent fuel elements must be chemically reprocessed; such reprocessing is therefore a late step. Again, for commercial feasibility, the power plant capacity must be several thousand megawatts. The number of commercial reprocessing plants in operation today is still very small. Their technological feasibility has been demonstrated in the past, even though the scale of these demonstrations was limited. If reprocessing is to be done on a very large scale, the feedback into the ecosphere will also be very large. This, in my judgement, will lead to an extension of today's technology. This extension does not appear to be a problem of technological feasibility: it is a matter of effort and of funds.

If chemical reprocessing is a late step, the final stage of waste disposal is an even later one. If very large-scale reprocessing probably requires an extension of today's technology, this – to say the least – also holds for the technology of the disposal of final waste. Disposal in adequately selected salt mines seems a suitable approach to this problem, and demonstration plants are being operated today.[2]

The picture of the peaceful application of nuclear energy for the economically competitive production of electricity is, therefore, today rather a simple one. LWR nuclear power plants are commercially feasible and are being installed and operated on a large scale. Those parts of the nuclear fuel cycle which must be carried out prior to, or at the same time as, the start of the operation of power plants are now commercially feasible also. Those parts of the nuclear fuel cycle which come into the picture only after the combustion of fuel elements require, in my judgement, an extension of present-day technology. The later they come into the process, the more they require the extension; such an extension is feasible. If, in the past, the nuclear power plant was the main target for research and development, in the future it will be that part of the fuel cycle which, in time, is last.

A similarly positive assessment can be made for commercial ships operating under nuclear power. The United States, the Soviet Union, Germany, and, soon also, Japan have or will have demonstrated the successful operation of such ships.

Only 10 years ago the picture was by no means as clear. There was an overwhelmingly large variety of competing types of nuclear power plants and the various features of the fuel cycle were not so readily discernible. There are now essentially two types of nuclear power plants which have evolved as second generation counterparts of the LWR. These are the high temperature, gas-cooled reactor, and the fast breeder reactor. I need not elaborate on the technical details, but will observe only that high temperature, gas-cooled reactors are capable of providing nuclear power at temperatures of 1,000°C. and more. This has certain advantages if electricity is to be produced. But the main thing is that, by virtue of high temperatures, nuclear power becomes applicable as chemical process heat. This means an extension of nuclear power beyond the production of electricity. In order to appreciate this feature fully one must realise that only roughly 25 per cent of the total demand for primary energy is a demand for the production of electricity. The remaining 75 per cent go in roughly equal parts into the sectors of transportation, domestic, and other industrial uses. The advantage of the high temperature, gas-cooled reactor is that it can make nuclear power applicable to these sectors.

The fast breeder reactor places the provision of fuel for nuclear power on a completely different basis. Fast breeder reactors extract on average roughly 100 times more energy from a given amount of natural uranium than can other nuclear reactors, including the high temperature, gas-cooled reactor. In the short term, this results in a remarkable stability in the price of uranium ore. At present ore prices, only one thousandth of the cost of generating electricity arises from the provision of ores. As a result, almost any uranium price can be afforded. This implies that even those ores which are of average richness can be processed. In the long run, this leads to a situation where nuclear power is capable of providing enough energy for several hundred thousand, if not million, years. The LWR of today can do that only for some decades. Both the high temperature reactor and the fast breeder complement the LWR remarkably well, and strategies can be identified which will ease the transition from market situations in the present day to world-wide problems of energy demand and supply in the future.[3]

In discussing the present position of nuclear energy, the element of time must be emphasised. The development of nuclear energy originally had a highly complex and sometimes disturbing appearance. Consideration of the temporal aspect led to a process of selection, in the course of which the valid parts of this development became evident. Those who were involved had to be educated and again this required time. For instance, it was through a step-by-step process that the electricity-producing firms – and national laboratories too – came to understand how to act in the development of nuclear power. Over and above all this, there is an inherent temporal dimension to such complex technological and organisational developments.

Large and complex developments have to pass over three thresholds. The first of these is the threshold of scientific feasibility: in the development of nuclear power for peaceful applications, the principal step towards scientific feasibility was taken by Enrico Fermi as early as 1943. The definitive consolidation of these steps occurred between 1943 and 1950. The second threshold is that of technological and industrial feasibility: to pass this threshold is a step the size of which is often underestimated. In the development of light water reactors in the United States this step comprised the successful establishment and operation of the Dresden, Yankee, and Indian Point – roughly 200 MWe – demonstration plants; in Western Germany, the plants at Obrigenheim and Lingen performed this function. The significance of operating these demonstration plants only began to be clearly evident in the early 1960s. In other words, it required nearly two decades to pass from the threshold of scientific feasibility to the threshold of technological and industrial feasibility. Fast breeder reactors and high temperature, gas-cooled reactors are passing the latter threshold only now.

The third threshold is that of economic and commercial feasibility. As a result of its complexity. This appears to be the most difficult threshold to those who have lived through the various stages of such a development. Let us recall that there were quite a number of reactor types which were technologically and industrially feasible but which were unable to pass over the third threshold. This is true for instance of the gas-cooled, graphite-moderated, natural uranium reactors which were developed in the United Kingdom and France. Economic and commercial feasibility implies that industry is able to sell its product at competitive prices without public subsidy. It also requires that the utility companies must be able and willing to place orders necessary for such very large-scale industrial projects and that the governmental bodies must have taken the necessary regulatory measures. It is this last step which turns out to be of outstanding importance. In the United States and, for instance, in Western Germany, acceptance of their appropriate roles and responsibilities by the various partners involved has been a long and sometimes complicated process.

Even now, about 30 years after Enrico Fermi's historic accomplishment, the process has not come to an end. The larger public has recently begun to realise the dimensions of this development. It is assuming a role in the third stage of the development which it did not accept in the first two phases. The objections to nuclear power which have been raised by certain parts of the larger public must therefore be taken seriously and fully understood.

Objections to Peaceful Nuclear Energy

These objections have arisen during recent years not only in the United States but also in Western Germany, Sweden, Switzerland, the Netherlands and Japan. It may be observed in passing that such opposition has not taken place – or at least not to such an extent – in Canada, the United Kingdom, France, Italy, and certain other countries. Generally speaking, the main objections to peaceful nuclear power appear to be the following: (1) the routine operation of nuclear power plants creates dangers of radiation; (2) if nuclear power is produced on a large scale, it is impossible to exclude the risk of a major accident, resulting in even greater dangers of radiation; (3) since fissile material can be used for both peaceful and non-peaceful purposes, there is a risk, in handling large amounts of fissile material, of the illegal diversion of such fissile material; (4) the operation of nuclear power plants necessitates the storage and disposal of radioactive wastes for periods of time which exceed human experience; (5) the large-scale operation of nuclear power plants involves the handling of large amounts of plutonium which must be kept out of the ecosphere; (6) the operation of nuclear power plants results in the production of large amounts of waste heat; (7) the erection of nuclear power plants requires too much land; and (8) the demand for more energy which is to be met by nuclear power is not real but is a product of manipulation.

These objections must be met. Many writers have been doing this over the past few years,[4] but this is not the task to which I am addressing myself in this paper. Rather, I wish to reflect on these arguments and to consider them in new categories, which will allow for a better understanding not only of the objections but also of the answers. To this end, I will consider more specifically a few of the objections.

The Routine Operation of Nuclear Power Plants: Today's routine operation of nuclear power plants leads to a radiation danger which is smaller than 5m rem/year. Recently it has also been made mandatory for LWRs to make the radiation danger smaller than 5m rem/year; the regulation said "as low as practicable". The question now is whether such a limit is acceptable. Originally the regulatory limit was 170m rem/year, but the practice was not to exceed 5m rem/year. A large margin therefore evolved. Dr. Gofman and Dr. Tamplin then made it a point to argue that not only the practical value but the regulatory limit as well should be correspondingly low. Their argument was as follows: there is experimental evidence to show that applying 1 rem induces about 20 additional cases of leukaemia per million persons. The relevant experiments were made at about 50 rem or more. If all kinds of cancer besides leukaemia are taken into account, the figure may be as high as 120 persons per rem per million persons. Then they calculated the number of cases of cancer by straightforward algebra, considering only 0.17 rem/year (i.e., 170m rem/year) and 200 million persons (equalling the population of the United States). This leads to $17 \cdot 10^{-2} \times 2 \cdot 10^{-8} \times 120 \times 10^{-6} \sim 4{,}000$ additional cases of cancer. Their conclusion was that this figure is too high.[5] A tremendous public debate resulted with much talk about the effects of poisonous nuclear power. It also led to new regulations by which a formal interpretation of the wording "as low as practicable" was required. The regulation now provides that the radiation danger be smaller than 5m rem/year.[6] This figure leads to 120 cases instead of 4,000.

The question necessarily arises: why is that the right figure? There is of course no answer to this. One cannot argue in the domain of absolute figures – comparisons by means of a yardstick are the only reasonable procedure. Such a yardstick can only come from placing the problem in the setting of nature. The "natural" radiation is, on average, 120m rem/year. The variation of that figure is high. Changing positions on the surface of the earth can easily lead to an increase of 50m rem/year. This is the case if one moves, for example, from Pittsburgh to the Smoky Mountains in the United States. Similarly, living in a concrete building leads to an increase of roughly 20m rem/year. In view of both, the average value and the variation of the natural radiation, 5m rem/year is a small figure. "Embedding" the problem into the normal conditions of life therefore does provide a yardstick. Similarly, one should make the comparison with the number of cancer cases which ordinarily occur. In the United States, there are 300,000 cases per year, so it is this figure against which the additional number of 4,000, or 120, as the case may be, must be assessed.

Many of the opponents of nuclear energy now say that statistical considerations are inhuman, and that any additional life which is "willingly" sacrificed is too much. Human beings have names. But this argument, by virtue of the simple algebra given above, implies a causal relation between the additional 5m rem/year and the additional 120 cases of cancer. While experimental evidence for this causal relation can be demonstrated, at the level of 50 rem, i.e., the ten-thousandfold value, this is certainly not the case at 5m rem. It is a problem of the extrapolation to low rates of dosage. Such an extrapolation is very hypothetical because there is no way of proving it experimentally. Further, such re-extrapolation requires that all other parameters be held constant. There is no individual, with a name, for whom this can be done. It is obvious that entering the hypothetical domain leads into strenuous debates which become so animated because they can never be resolved. "Embedding" is the way out of this dilemma. The normal conditions of life provide a yardstick or standard.

The International Committee on Radiological Protection (ICRP) deals with standards by establishing limits of radiological dosage on an international basis. This requires understanding the causal relations between rates of dosage and damage. Understanding these causal relations, however, is only one side of the coin. The other side of the coin is the standard of what is acceptable. This, too, is done through the process of "embedding". The ICRP shows it is aware of this by making the following observation:

Any exposure to radiation is assumed to entail a risk of deleterious effects. However, unless man wished to dispense with activities involving exposures to ionizing radiations, he must recognize that there is a degree of risk and must limit the radiation dose to a level at which the assumed risk is deemed to be acceptable to the individual and to society in view of the benefits derived from such activities.[7]

"Embedding" entails the establishment of a criterion which incorporates both of the actions recommended – in this case, the erection and operation of nuclear power plants – and the benefits and risks of the alternatives. The most straightforward way of doing this is to consider the ratio. Up to now, the ICRP has refrained from identifying the denominator of that ratio, the benefit, on the grounds that the benefit varies from society to society and from one situation to another. The assessment of radiation standards in a situation of oil shortage, for instance, is one thing; considering such standards in a situation where large amounts of very inexpensive energy exist is another. In addition, benefits may appear differently to poor countries and rich ones. For these reasons, the ICRP decided to leave the evaluation of such benefits, and therefore the ratios of benefits to risks, to the respective national bodies.

There is a further point: the ICRP talks of risks. There is little doubt that if the process of radiation damage were completely understood it would be possible to assess that radiation damage deterministically. Damage could be assessed with certainty. Risks of the kind we are discussing refer to events governed by laws of nature of which we have incomplete knowledge.[8] If our knowledge were complete, the ratio in question would become a ratio of benefit to damage. The problem would not be one of the acceptability of risks but of the acceptability of damages which are certain. In either case, it is acceptability which is at issue.

Comparing risks – or damage – arising from a proposed activity with those occurring in nature is one way of dealing with the problem of acceptability. Another way is to ask for alternatives to the activity proposed. For instance, are there alternative possibilities of producing sufficient amounts of energy at prices which purchasers are willing to pay? Such alternatives are features of the world we live in and this also might therefore be viewed as "embedding". For some time to come the possibility of energy production by the combustion of fossil resources will remain. For the moment we will leave aside the problem of the adequacy of resources. The point here is rather that the use of fossil fuel is also accompanied by pollutants which damage health. The most notable is SO_2. Of course a Gofman-Tamplin analysis can be made here too. It has been reported that for every 100 persons-ppm-year, there is roughly one death which is a result of respiratory infection.[9] In larger cities, SO_2 concentrations in the order of 100 $\mu g/m^3$ of SO_2 can often be observed. This relates to the order of 0.1 ppm. Making the assumption here that about 60 million persons in the United States live in areas of such SO_2 content, this results in $6 \cdot 10^7 \cdot 0.1 \cdot 10^{-2} = 60,000$ deaths/year as a result of SO_2. The problem raised by Drs. Gofman and Tamplin is therefore not peculiar to nuclear energy. It can always be raised when low rates of dosage of any toxin or drug are considered. Needless to say, in the case of SO_2, as in that of radiation, the question of the causal relationship must be considered. Again we enter the domain of the hypothetical.

For each of the various pollutants, such as SO_2, a standard must be established; the analogous problem of the establishment of radiological standards must then be treated in the same manner. This immediately raises the problem of how we can ensure that all the standards in question are equally rigorous. The question necessitates a comparison of risks – or damage – of completely different kinds. How does pollution by SO_2 add to that of dust, and does the sum of both compare to 5 mrem/year? The difficulty is, however, broader than that. How is a dusty atmosphere reiated to a risk of having a 1/2 per cent increase in the rate of mutation within a few generations? The time element complicates the issue even further.

Up to now the approach to these questions has been implicit or pragmatic. In the past this was satisfactory because the related issues were of limited size and therefore of limited significance. But in the life of coming generations, with the increased use of technology which produces more pollution and the prospect of a greatly increased population and a greater demand for energy, this becomes a question of much greater urgency. It is necessary therefore to look for methods to approach these problems in a more coherent and systematic way. Furthermore, the interest of mathematicians in these problems has helped to make for a greater awareness of the possibility of treating the problems more systematically through the use of concepts like the value problem, vector, value optimisation, the decision process, and others.[10] The problem of the evaluation of risk raises basic methodological problems. Only very little knowledge is available,[11] and much work is required in this field.

The task may be formulated in summary form along the following lines: it is necessary to establish standards for the impacts of certain technological measures. In doing this we will contribute to the evaluation of related problems. The establishment of standards entails relating, as thoroughly as possible, ambient rates of dosage to physiological and other damage. At the same time, we must establish criteria and rules of acceptability, including both damage which is certain and the risk of damage. This requires "embedding" the standards and criteria into the normal conditions of life. For the evaluation of alternatives, it is necessary to establish standards of a completely different nature. This inherently requires their comparison and involves a process of decision, i.e. a problem of value.

Rare but Major Accidents: Nuclear power plants are made to be safe on the basis of passive or active engineering measures. This means that engineering measures either act by their mere existence – for example, as shields and locks – or by positive action – for example, as moveable control rods. Any such engineering measure can fail and in an advanced state of technology it is possible to assess a probability of such failures. This has been done in electronics in particular. While small, such probabilities of failure are nevertheless finite. If the probability of failure is considered to be too high, a second line of defence has to be installed. Safety components must be installed which are redundant. By the successive staging of such safety devices, it is possible to make the probability of failure smaller than any given small number[12]; but it is not possible to reduce it completely to zero. This, then, raises a number of problems. The first of such problems was stated by Dr. Starr as "How safe is 'safe enough'?" In recent years, the concept of failure rates has been introduced. A rate of failure of, for example, 10^{-x} per year, might be considered. For instance, x could be between 3 and 7. This leads into the intricate problem, which is partly a semantic one: is a plant of a failure rate of 10^{-7} really 10 times safer than a plant of 10^{-6}? If the term "safe" is clearly defined by such failure rates, this is certainly so. But is this definition adequate? What is the more general meaning of "safe" to a wider public? And can such a more general meaning really be quantified?

When we ask "How safe is 'safe enough'?" we are again led into the process of "embedding". Other risks have to be considered and alternatives have to be evaluated. Solar power and geothermal energy resources must be considered and compared with nuclear power. These alternatives have their own risks. If the residual risk of operating a nuclear power plant is to be compared with these risks, one has to reflect on the fact that the residual risk in the case of nuclear power is not a risk to be considered without reference to alternatives. There is little doubt that the laws of nature which are employed in the design and operation of, for example, a control rod are quite adequately understood. But, for the application of these laws of nature, knowledge of all initial and boundary conditions is also required. These initial and boundary conditions are not part of the laws of nature but are required if the laws of nature are to be applied to a particular case. This is necessarily so. The generality of the laws of nature results from the abstraction of the elements of the actual case, which are the initial and boundary conditions. As an example, let us think of the differential equation for a mechanical movement. Newton's law is general and governs all mechanical movements. But if the fall of a particular apple from a particular tree is to be predicted, the locus and the momentum of the apple at a given moment in time must be provided.

Measurement is the only way to accomplish this. This is a principal observation. The difference between the nature of initial and boundary conditions and the laws of nature has been dealt with recently by Professors Carl Friedrich von Weizsäcker and Erhard Scheibe,[13] who have described the nature of initial and boundary conditions as contingent. In so doing, they employ the medieval philosophical term contingere, which is a translation by Boethius of the more basic Greek word ἐνδέχομαι. It means encountering an event without being able to predict it; it simply happens. In this context, my usage of the word "contingent" is different from the common English word "contingent", which has a somewhat different meaning but which has the same root. Leaving aside the still more basic problem of interpretations in quantum theory, one must make the further observation that it is impossible to measure initial and boundary conditions with the completeness and accuracy necessary for a fully deterministic prediction of the performance of a technical component or device. This is where risk resides. It is not a risk of the type which is related to uncertainties in the knowledge of the employment of the laws of nature. It is a risk of the kind which refers to incomplete knowledge of initial and boundary conditions. While, at least in principle, it is possible to eliminate risks of the first kind, there is no possibility of doing so with risks of the latter kind. We can always improve our knowledge about contingent elements, but we can never make it complete. This restates the proposition that the residual risk can be made smaller than any given small number but it cannot be reduced to zero.

The traditional engineering approach to eliminating risks of the second kind – the risks which are integral to contingency – is trial and error. The engineer learns by experience to make better and safer machines. This is close to the scientific approach: an hypothesis is made which is followed by experiments, which in turn lead to an improved hypothesis, which again is followed by experiments. In this way a theory evolves which is true, i.e., is in touch with reality: *Veritas est adequatio rei et intellectus.*[14] It is precisely this interplay between theory and experiment, or trial and error, which is no longer possible for new technologies which are designed to master unique challenges. In the case of reactor safety this is obvious. It is not acceptable to learn reactor safety by trial and error. Before we continue this line of reasoning, another observation is in order.

Reactor engineers face this dilemma by dividing the problem of engineered reactor safety into sub-problems. For instance, the integrity of a pressure vessel is investigated as a sub-problem, and so is the operability of control rods and pumps. To use the performance of such components to draw conclusions about the performance of the whole nuclear plant of course requires a certain preconceived idea of the various interactions which must be considered when the components are combined. In combining such components more contingent elements come into the picture. The aim is to minimise the impact of incomplete knowledge of contingent elements. Therefore, in designing the needed experiments – which are tests, in this context – the largest possible units are sought. But, by the same token, the generality of the conclusion is reduced just because it was the contingent elements which led to these tests of large technological units. In persisting further and further along this line of reasoning, one arrives at a situation where the truly large-scale test can only result in the statement that a given device functioned at a particular time and place. A general conclusion is impossible. Because of the fact that the test stressed the contingent aspect of the elements, all generality is eliminated. Reactor engineers are familiar with this feature of reality. To some extent, the term "integral test" refers to it. We should recall the debate on the advisability of reactor tests of the Borax type,[15] where the argument was raised that almost nothing general could be concluded from such an integral test. Only large instrumentation could justify such an expensive test. But this refers exactly to the measurement of contingent elements. At the extreme, one may call such large-scale integral tests "happenings".

Let us again return to the general line of the reasoning followed in this paper. Reactor safety in its ultimate meaning cannot be evaluated by trial and error. Subdividing the problem can lead only to an approximation to ultimate safety. The risk can be made smaller than any small but predetermined number which is larger than zero. The remaining "residual risk" opens the door into the domain of "hypotheticality". At this stage, it becomes clearer that the concept "hypotheticality" is crucial in this analysis. A few words of explanation are therefore necessary. Hypotheticality, of course, is not a word in regular usage but its logic expresses precisely what must be expressed in the line of reasoning presented here. Its logic is the same as that of the word "criticality", for example, a term which is familiar to reactor engineers. The rule followed is that for Latin words ending in -itas, for example, veritas or felicitas. Such substantives point to features which exist in principle and which if actualised, lead to the fact that something can have a certain property: a reactor can become critical or a situation can be considered to be hypothetical. The process of iteration between theory and experiment which leads to truth in its traditional sense is no longer possible. Such truth can no longer be fully experienced. This means that arguments in the hypothetical domain necessarily and ultimately remain inconclusive. I think that this ultimate inconclusiveness which is inherent in our task explains, to some extent, the peculiarities of the public debate on nuclear reactor safety. The strange and often unreal features of that debate, in my judgement, are connected with the "hypotheticality" of the domain below the level of the residual risk.

Why is it impossible to apply the method of trial and error to ultimate reactors safety? It is impossible because the consequences of so doing would be too far-reaching. Every country is too small for that – eventually even the globe itself is too small. The magnitude of the technological implications thus becomes comparable with the magnitude of the constraints which determine our normal life. This inapplicability of the scheme of trial and error may be regarded as a kind of cost – or risk. But nuclear power is to be justified by its capacity to provide practically infinite amounts of energy for even a highly populated and industrialised world. This is an unusual kind of benefit which exceeds any so far experienced. The appropriate concern therefore is whether the ratio of such costs and benefits is adequate. To complete the picture, the magnitude of engineered safeguards also is inherently very large, and is hitherto unexperienced. The inverse failure rates which relate to the periods of time within which to wait for a failure are an indication of that. The magnitude of the constraints of the world we live in corresponds, if properly interpreted, with the magnitude of the costs – or risks – the magnitude of the engineering measures needed to guard against it, and the magnitude of the benefits.

In dealing with reactor technology, we should realise that nothing less than this is at stake. I think that it is insufficient to look at nuclear power plants as mere substitutes for the present more conventional power plants which use fossil fuels.[16] In line with this reasoning, it is perfectly legitimate to look for alternatives to nuclear power. These alternatives must, however, be evaluated by means of the same global standards or within the same magnitude and scope as nuclear power. To give an example: Tidal power is not an alternative. It cannot provide the same amount of power as can nuclear power.[17] If nuclear power proves superior to other alternatives when properly evaluated, it will be fully legitimate to proceed with what will then be the truly large-scale development of nuclear power.

Let us for a moment consider some of these choices. As mentioned before, one such alternative is solar power. At first inspection, solar power appears to be a clean source of energy. Solar power has, however, the disadvantage of being dispersed; one cannot hope to harvest more than, say, $20W/m^2$. Large areas of the globe must therefore be utilised to provide very large amounts of solar energy. The present consumption of energy in the world is at $8 \cdot 10^{12}W$; in the future, $60\text{-}150 \cdot 10^{12}W$ must be prepared for, as the world population increases and the poorer countries begin to obtain their share. A figure of $100 \cdot 10^{12}W$ at $20W/m^2$ implies a figure of $5 \cdot 10^{12}$ m². This is about 20 times the area of the German Federal Republic. To take such a step might not be impossible eventually but its size and nature imply a kind of impact on the normal conditions of life – including the political aspects of such normal conditions – which indeed offers little advantage over the impact of other alternatives. If produced on a very large scale, solar power would not offer a solution which left the normal conditions of life unchanged.

There are other repercussions of the development of solar power besides the use of very large tracts of land. The energy balance of the atmosphere above these large areas would of course change accordingly. One has to realise that such changes would be likely to happen in stages. The most probable first stage would be a change in the pattern of rainfall, but not in the global average of the amount of rainfall. Even if nothing else were to happen, there would be dramatic consequences: the continuous drought in North Africa which we experience today is indicative of that. In the next stage, the average amount of rainfall and its pattern would alter together with regional changes in climate. One result might be a regional change of average temperatures and winds. In the third stage, the global climate would change with a change in the average global temperature. This could cause an ice-age or melting of ice-caps. It is not good enough to point out that even large-scale harvesting of solar energy would make up only 1 per cent. or so of the total solar input and might therefore be negligible. The question is more delicate than this since regional instabilities of weather and climate are at stake. It is equally inadequate to point out that regional uses of solar power are desirable and clean in many if not all respects. Only if solar power can be produced on a very large scale can it be considered to be a legitimate alternative to nuclear power. It may very well be the case that eventually a combination of solar, nuclear and possibly even other additional forms of energy might turn out to be the optimal condition. I mentioned earlier that a major effort is now under way to develop comprehensive computer programmes for the largest computers available, and to develop appropriate global systems for monitoring weather conditions, pollutants, and other parameters such as the CO_2 content of the air.[18] This will bring significant progress. But it will always be impossible to run truly integral experiments as they were discussed above in the case of nuclear reactor safety. The prediction of man's impact on the climate will never be fully proved experimentally, or, to be more exact, by trial and error. The magnitude of the risk which is involved again corresponds to the magnitude of the constraints which determine our normal life. A residual risk will therefore remain. Again one finds oneself in the domain of "hypotheticality" and the debate on the related issues will be inconclusive.

It is now clear that a similar argument holds for the very large-scale "harvesting" of the heat in the earth's crust. This statement implies, among other things, that the risk of inducing large-scale earthquakes is unacceptable. Furthermore, contrary to a widespread belief, the fusion reactors which are hoped for would by no means be a source of clean power.[19] More systems-analyses comparing various alternatives for the provision of energy are now urgent, although it is unlikely that one of the alternatives would turn out to be clearly superior to all others.

The ecological impact of human activities is also of that nature. Once ecological equilibria are destroyed, it takes periods of time beyond human experience for their re-establishment. In this case, too, trial and error, or, rather, the iteration of theory and experiment, cannot provide a definitive solution. Residual risks will remain.

Many more examples of this situation can be found. The tasks which we confront in the assessment of nuclear power are prototypical of a wider range of emerging problems; in this sense, the analysis of the problems of nuclear power is a pathfinding undertaking. If properly interpreted and understood, the public concern about nuclear power is not unfounded. But that concern is not a simple function of a peculiarity of nuclear power. It is, rather, the general condition of civilisation towards which we are moving;

it is a condition where the magnitude of human enterprises becomes comparable with the magnitude of the widest determinants of our normal existence. Nuclear power turns out to be a forerunner, a pathfinder, of that.

Nuclear Material Safeguards: Another objection raised by the opponents of nuclear power asserts that it will render easier the illegal diversion of nuclear material for military purposes. This is sometimes called the problem of safeguards. A vast international effort has been made to deal with the problem. The conception, negotiation and commencement of the Treaty on the Non-Proliferation of Nuclear Weapons marked a great step forward in this respect. A major effort of systems-analysis and development led to the establishment of the safeguards system of the International Atomic Energy Agency (IAEA). The continued operation of the European Atomic Energy Community as well as the United States Atomic Energy Commission safeguard systems have also contributed to a considerable reduction of the risk of illegal diversion.

As in other situations in which risks of catastrophic dimensions must be dealt with, so in the case of safeguards also the establishment of standards was in the forefront of interest. In order to establish thresholds of significance, it was necessary to obtain experimental values for the material unaccounted for (MUF), since the accountability of nuclear material is the corner-stone of the present safeguard systems. The side of the task which deals with causal relations of a scientific or technological nature was more or less readily taken care of. To that end a number of experiments were designed and executed. The result was that the amount of MUF for the various nuclear facilities such as reprocessing plants was established. On the other side, it was necessary to reflect on the demands for such accountability. Was the IAEA system of safeguards to be designed to account for kilogrammes, grammes, or milligrammes of nuclear material? And what were the periods of time during which the detection of diversion had to take place? It was a painful and lengthy process to explain to politicians that it would be impossible to account for absolutely all the nuclear material. A residual percentage would go into MUF; there was bound to be a residual risk in the case of safeguards, as in other situations of the kind under discussion.

What then is acceptable? In the case of international safeguards against the illegal diversion of nuclear material, the need to accept residual risks has led to highly hypothetical considerations. One such hypothetical problem was whether it would be considered feasible that someone should drill little holes into thick concrete walls for the diversion of x grammes of uranium. The domain of the hypothetical comes even more clearly into focus if I report here on the concern that there might be someone in a hidden back room in the basement of a power plant preparing plutonium without paying attention to any radiation this might cause. As a result of an intensive international effort under the guidance of IAEA for several years, it was eventually possible to narrow the gap between the demand for international safeguards and their feasibility. It was possible to assess the acceptability of residual risks. Yet absolute certainty of the safeguards cannot be provided; a residual risk remains and therefore the situation is "open-ended".

The responses which can be given to those who object to the development and use of nuclear power are pretty much the same for all the arguments. There is one exception. This is the argument which asserts that the demand for energy is the result of manipulation and is not "real". It is an interesting question, but it does not involve scientific or technical considerations of the kind which I have been treating in this paper. Hence, I shall not deal with it here.

Marginal Remarks on the Energy Problem

It nevertheless seems appropriate to make a few remarks on the nature of the problem of energy which industrial countries are facing, as this establishes the scope for any systems-analytical evaluation or assessment of nuclear energy or related phenomena. It is most important to realise that the problem of energy appears in phases over time, the features of which are sometimes very different. In the first phase of the problem, it is the oil shortage which is to the forefront. This phase will last for at least 15 years because any technological measure requires that much time before its impact attains substantial economic significance. During this phase, research and development on the substitution of synthetic hydrocarbon fuels for oil are urgently needed. The second phase is likely to be characterised by the large-scale use of nuclear power in the production of electricity and of coal for the non-electrical sector of demand for secondary energy. This phase may last for a few decades.

In the long run, i.e., in the third phase, all demand for primary energy must be met by other than fossil fuels. There are four alternatives: nuclear fission energy, nuclear fusion energy, solar power, and geothermal energy from the earth's crust. All four alternatives can provide very large amounts of energy, although in different degrees.

During the second phase, preparations for changing to one of these alternatives or to a combination of them must be made. These preparations will require more than the exclusive consideration of the supply of fuel. The procedure of "embedding" is needed in order to arrive at an optimal solution. The problem of fuel supply must be "embedded" in the atmosphere, the hydrosphere, the ecological sphere and the sphere of social life. The demand for and supply of energy and their embedding have to be considered on a scale which might be something like 25 times larger than today's scale. It must be assumed that the world's population will increase and that the poor countries will begin to approximate the consumption of energy *per caput* which is close to that of the rich countries.

This, then, is the setting for assessing the various technological alternatives for providing enough energy and their various advantages and disadvantages. The promulgation and comparisons of standards, the assessment of residual risks, "hypotheticality" and "embedding", are the intellectual procedures which must accompany the technological developments of tomorrow.[21]

"Hypotheticality" and Arms Control

It is probable that "hypotheticality" will characterise the next stage of human enterprise. The magnitude of technological enterprises will be so great that it will not be possible to proceed with the absolute certainty that there will be no negative consequences. The magnitude of the undertaking and the intellectual impossibility of eradicating all contingency make the prospect for "hypotheticality" even more certain. But if we are condemned to "hypotheticality", are there ways to deal with it? Most of these ways are still to be explored. The case of arms control, however, might illustrate the way of proceeding which is available to us. It was recognised in the case of nuclear armament earlier than in other spheres that it was leading into qualitatively new conditions. The yield of atomic weapons exceeded all previous experience and the means of delivering such weapons permitted delivery from almost any part of the earth's surface to almost any other part. The magnitude of the military and technological implications became comparable with the magnitude of the determinants which set the framework of our normal existence. Furthermore, it is impossible to develop the art of nuclear warfare by trial and error, as was previously always the case with innovations in military technology. Hence, residual risks became evident in the field of atomic warfare. This happened early and a number of schools of thought have evolved to deal with these residual risks and to develop a practicable rationale or doctrine.[22]

The kind of reasoning which developed in these schools of thought soon became very specialised. It was not easily possible for a citizen who was not a scientist, or even for scientists who were not working intensively in this field, to follow these developments. The debate on anti-ballistic missile systems (ABM), on multiple independently targeted re-entry vehicles (MIRV), and other such problems, demonstrated this. The reasoning involved in these decisions was extremely complicated and abstract. It was also extraordinarily difficult to propound absolutely conclusive arguments. There was always another view and there was no way of ultimately proving one's own position. At the same time, the whole domain of nuclear armament was vitally urgent. It was considered to be a matter of life and death to have a strategic doctrine which would deter one's visible and rational opponent from engaging in nuclear warfare. This feature of being extremely unreal – abstract and irrelevant – and at the same time extremely real – and relevant – developed into a serious gap in the late 1960s when the ABMs and MIRVs came into the picture. At that time SALT negotiations started. For a long period it was not clear whether a treaty would be agreed to. But even if they had not resulted in such a treaty, it would indeed have been an achievement to have held the SALT negotiations. In the course of these negotiations the two antagonistic parties talked to each other and came to understand better each other's rationale of nuclear strategy. It was a matter of enhancing the prospect of the survival of the human race. It gave a hint about how to deal with "hypotheticality".

This simultaneity of two contradictory modes of appearance is well known in the rigorous interpretations of the meaning of quantum theory. In order to overcome contradictions such as that between wave and particle modes of appearance, it was necessary to have recourse to more pronounced abstractions. Thus, it was no longer feasible to describe the laws of quantum theory – and hence the laws of nature for that matter – in the ordinary space-time domain. Only the abstract Hilbert space was feasible for the description of the laws of nature. At this level of abstraction, there was no longer a contradiction between the two modes of appearance. But the translation of statements which belong to Hilbert space in the ordinary space-time domain always leads to some kind of strange phenomena, like the principle of uncertainty, which are difficult to cope with if such abstraction is not employed.

Nuclear warfare and arms control were the first human activities which touched upon the constraints and conditions which are imposed on us by living on our finite earth. The rational treatment of nuclear warfare and arms control leads immediately into the domain of "hypotheticality" with all its peculiarities. Nuclear warfare and arms control not only touch the finite limits of the earth; they bear very intimately on our interdependent large-scale civilisation in which there are no isolated areas for instructive trial and error.

The Formalised Debate

Many of the statements and actions of the opponents of nuclear energy are wrong and even harmful. It is by no means the purpose of this article to imply that they are right. They are not. But they have led to debate and hence some of their effects must be estimated positively. These debates have positive consequences. The regulatory interpretation of the phrase "as low as practicable" in the case of LWRs is one such positive consequence. To have aroused a more acute awareness of the potential – and hypothetical – dangers is another positive result. Such results depend upon antagonists talking to each other and listening to what the other has to say. For this to happen on a large scale is not easy. But in a democratic society it must occur on a large scale because, eventually, decisions about acceptability profoundly affect every member of society. These issues are beyond the powers of science but they are of vital concern to citizens. Dr. Alvin Weinberg has called the domain of these issues "transscience".[23] To facilitate a general debate on these issues a high level of formalisation appears to be necessary. Abstract and complex problems beyond human experience or the experience of everyday life must be dealt with.

The licensing process which is required for the construction and operation of nuclear power plants in the United States has already established such a formalised debate. The necessity of taking a position under the pressure of deadlines is very conducive to hard thought. Design engineers, including myself, have tried to come to such conclusions before the formal licensing procedure by strenuous thinking and by prior consultations with the individual members of the licensing bodies. However, these methods have not been satisfactory. Only the formalised confrontation with the antagonist's arguments, when the licensing body is acting in full responsibility, permits the kind of analysis needed for a design in a particular case.

The treatment of particular cases by a formalised procedure for dealing with particular situations forces abstract considerations to take concrete form. All this imposes time-limits and enforces specificity and concreteness in analysis. The licensing process takes place in several rounds. It is an iterative process which usually produces convergence from diverse positions. This iterative process in time supplants the iterative scheme of trial and error: *Veritas est adequatio rei ad intellectum*.[24] It bridges the gap which is imposed on us by "hypotheticality". But formalisation goes even further. It allows a great number of parties to state their case. In a formalised debate, this accentuates the iterative process. This also tends to bridge the gap of "hypotheticality".

In this setting too, nuclear energy, by leading to this kind of a licensing process, has turned out to perform a pathfinding function. Formal licensing procedures using public hearings are now being applied to situations other than those involving nuclear energy. If in the previous paragraph on arms control it became apparent that it is the element of debate which is necessary and, further, that a stride towards more abstraction is required if one wants to meet present and future challenges, now, following the observations of this last paragraph, it seems possible to go on a step further and to identify the object of such abstraction. It is the debate itself. Procedures must be devised which permit the intangible to be made tangible. A rationale for the formalisation of the debate here envisaged must be established.[25] It appears that in this respect too nuclear energy serves as a pathfinder.

66

[1] Häfele, Wolf and Seetzen, Jürgen, "Prioritäten der Grossforschung" in Grossner, Claus, *et al.* (eds.), *Das 198. Jahrzehnt: Eine Team-Prognose für 1970 bis 1980* (Hamburg: Christian Wegener Verlag, 1969), pp. 407-435; Servan-Schreiber, Jean-Jacques, *Le Défi Américain* (Paris: Denoël, 1967).

[2] Wittenzellner, R., "Storage of Solid Radioactive Waste", and Tuohny, T., "Managing Liquid Radioactive Waste", discussions on the storage of nuclear waste, 20 September, 1973, at the 17th Regular Annual Session of the IAEA General Conference, Vienna, 18-24 September, 1973.

[3] Häfele, Wolf and Schikorr, Winfried, "Reactor Strategies and the Energy Crisis" IAEA Study Group on Reactor Strategy Calculations, Vienna, 5-9 November, 1973.

[4] See, for instance, "The Nuclear Controversy in the USA", International Workshop, Lucerne, Switzerland, 30 April-3 May, 1972, sponsored by the Swiss Association for Atomic Energy in cooperation with the United States Atomic Industrial Forum; and "Energie und Umwelt Informationsdienst der Zeitschrift", which appears in each issue of *Atomwirtschaft – Atomtechnik*.

[5] Actually, the figure which Drs. Gofman and Tamplin were talking about referred to 30,000 persons, because they employed a different set of medical data. But this does not influence the logic of the argument presented here. Gofman, J.W. and Tamplin, A.R., *Poisoned Power: The Case against Nuclear Power Plants (Emmäus, Pennsylvania: Rodall Press, 1971)*.

[6] United States Atomic Energy Commission, "Proposed Rule Making, Licensing of Production and Utilization Facilities. Light-Water-Cooled Nuclear Power Reactors. /10 CFR Part 50/", *Federal Register*, XXXVI, 111 (June 1971). See also, United States Atomic Energy Commission, "As Low as Practicable Numbers could end the Radiation Controversy...", *Nucleonics Week*, XII, 22 (June 1971), p. 4.

[7] ICRP, "Recommendations of the International Commission on Radiological Protections", *ICRP Publication*, pt. 9, A(34) (Oxford: Pergamon Press).

[8] This is a general observation. See the recent discussion of this problem in Brooks, Harvey, "Science and Trans-science", in Correspondence, *Minerva*, X, 3 (July 1972), pp. 484-486.

[9] Wilson, R., "Tax the Integrated Pollution Exposure", *Science*, CLXXVIII, 4057 (October 1972), pp. 182-183.

[10] Raiffa, Howard, *Decision Analysis: Introductory Lectures on Choices under Uncertainty* (Reading, Mass.: Addison-Wesley, 1968).

[11] See, for instance, Starr, Chauncey, "Benefit-Cost Studies in Socio-Technical Systems", Colloquium on Benefit-Risk Relationships for Decision-Making, Washington, D.C., 26-28 April, 1971, sponsored by the United States National Academy of Engineering. See also, Otway, Harry and Erdmann, Robert, "Reactor Siting and Design from a Risk Viewpoint", *Nuclear Engineering Design,* 13 (1970), p. 365.

[12] Häfele, Wolf, "Ergebnis und Sinn des SEFOR Experiments", in Scheibe, E. and Süssman, G. (eds.), *Einheit und Vielfalt, Festschrift für C.F. von Weizsäcker zum 60. Geburtstag,* (Göttingen: Vandenhoeck and Ruprecht, 1972), p. 248.

[13] von Weizsäcker, Carl F., "Komplementarität und Logik", *Die Naturwissenschaften,* XLII, 19 (1955), p. 521. See also Scheibe, Erhard, *Die kontingenten Aussagen in der Physik* (Frankfurt: Athenäum, 1964), and *The Logical Analysis of Quantum Mechanics* (Oxford: Pergamon Press, 1973).

[14] Aquinas, Thomas, *Quaestiones disputatae de veritate,* verit 1. 1c. This formula has been widely referred to in the history of medieval philosophy and stresses the objectivity of things.

[15] Dietrich, Joe R., "Experimental Determination of the Self-Regulation and Safety of Operating Water-Moderated Reactors", *Proceedings International Conference on Peaceful Uses of Atomic Energy,* August 1955, Geneva, United Nations, New York, vol. 13 (1956), pp. 88-101.

[16] Weinberg, Alvin, private communication.

[17] See, for instance, Hubbert, M.K., "The Energy Resources of the Earth", *Scientific American,* CCXXV, 3 (September 1971), p. 60.

[18] World Meteorological Organisation and International Council of Scientific Unions, "The First GARP Global Experiment, Objectives and Plans", *GARP Publication Series,* nr. 11 (March 1973).

[19] Häfele, Wolf and Starr, Chauncey, "A Perspective on Fusion and Fission Breeders", *Journal of the British Nuclear Energy Society (forthcoming).*

[20] International Atomic Energy Agency, "The Structure and Content of Agreements between the Agency and States Required in Connection with the Treaty on the Non-Proliferation of Nuclear Weapons", IAEA, Vienna. Information Circular 153, 1971. See also, Häfele, Wolf,

"Systems Analysis in Safeguards of Nuclear Material", *Proceedings Fourth International Conference on Peaceful Uses of Atomic Energy,* September 1971, Geneva, United Nations, New York, IAEA, Vienna, vol. 9 (1972), p. 303.

[21] Häfele, Wolf, "Energy Systems", in *Proceedings of the IIASA Planning Conference on Energy Systems,* Laxenburg, Austria, International Institute for Applied Systems Analysis, July 1973.

[22] A well-known example is contained in Wiesner, Jerome B., "Comprehensive Arms-Limitation Systems", in *Arms Control, Disarmament and National Security* (New York: George Braziller, Inc., 1961), pp. 198-233.

[23] Weinberg, Alvin, "Science and Trans-science", *Minerva,* X, 2 (April 1972), pp. 209-222.

[24] This formula and similar ones have been widely used in the history of medieval philosophy to stress things as objects of comprehension, in contrast to Thomas Aquinas' formula.

[25] To some extent this is concerned with the problem of the role of scientists as advisors. This problem has been examined in *Minerva,* X, 1 (January 1972), pp. 107-157; X, 2 (April 1972), pp. 280-294; X, 3 (July 1972), pp. 439-451; X, 4 (October 1972), pp. 603-613; XI, 1 (January 1973), pp. 95-112; and XI, 2 (April 1973), pp. 228-262.

However, the point which is developed here is not exactly the problem of scientific advice. It is rather the function of science in designing the rationale of a formalised debate. The partners in such a debate can be of many different kinds. Science is not necessarily the leading partner in this.

SECTION 2:
THE NUCLEAR OPTION

ENERGY OPTIONS AND THE ROLE OF NUCLEAR ENERGY IN ASIAN COUNTRIES

K.T. THOMAS

I propose in this paper to analyze energy production and consumption patterns in the Asian countries, their minimum needs for attaining a barely satisfactory standard of living, the gap between the two, and how alternate sources of energy including atomic energy have to be seriously considered in bridging this gap. The countries covered are:

1. Middle East including Egypt, Iran and Turkey
2. India
3. Pakistan, Burma, Sri Lanka
4. Indonesia, Singapore, Malaysia and the Philippines
5. Thailand, Laos, Khmer Republic, Republic of Vietnam
6. Republic of South Korea
7. Japan

These countries vary a great deal not only from the point of view of energy consumption. In a paper like this it is not possible to cover all the aspects in full, but one can only try to do a macro-analysis of the situation and paint the scenario with a broad brush.

Table I gives details of per capita national income in US dollars and per capita energy consumption in kgs. coal equivalent, of principal Asian countries compared with Europe and North

Table I
Energy Consumption Vs National Income

S.No.	Country	Per capita National Income (US dollars)	Per Capita Energy Consumption (kgs. coal equ.)
1.	Kuwait	3703 (1971)	10,441
2.	Japan	2462	3,251
3.	Israel	2007	2,712
4.	Singapore	1041 (1971)	885
5.	Lebanon	521 (1970)	889
6.	Malaysia	391	491
7.	Iran	367 (1971)	954
8.	Turkey	335 (1971)	564
9.	Syria	295 (1971)	455
10.	Republic of Korea	281	827
11.	Iraq	278 (1969)	642
12.	Jordan	276	331
13.	Philippines	254	311
14.	Saudi Arabia	*	900
15.	Pakistan	205 (1971)	158
16.	Egypt	202 (1970)	324
17.	Thailand	193	305
18.	Rep. of Vietnam	174 (1971)	287
19.	Sri Lanka	164 (1971)	146
20.	Khmer Republic	*	25
21.	Indonesia	112	133
22.	India	91	186
23.	Burma	68 (1969)	58
24.	Laos	*	79
25.	Bangladesh	–	32
26.	World	930 (1970)	1,984
27.	USA & Canada	5000	11,526
28.	Europe (except Eastern)	2740	4,000

Source: U.N. Statistical Yearbook 1973

All the figures unless otherwise indicated are for 1972.

* Figures are available for these countries only upto the year 1963. They are:

Saudi Arabia	230
Khmer Republic	117
Laos	61

America. The data have been collected from the latest information available and are applicable mostly to the years 1970-72. It is recognized that after the oil crisis in 1973, the situation has changed radically and therefore, the data may need considerable updating. The general analysis, however, will not show any appreciable change in trends. The per capita national income varies from a low US $ 68 (1969) for Burma to US $ 3700 (1971) for Kuwait, whereas the average figures of USA and Canada combined was US $ 5000 in 1972 with the world average at US $ 930 in 1970. The lowest figure was thus about 1/75th of the US per capita income, 1/40th of Western Europe and about 1/14th of the world average. The basis for considering whether a country is developed or not has been discussed on many occasions in the past but no universally accepted guidelines exist. The status of a country can be gauged by the standard of living of its people as compared with that which can be sustained by production based on the present level of science and technology. The per capita national income of the world is sometimes regarded as an appropriate dividing line to consider whether a country is developed or not. With the exception of Burma, India and possibly Laos and Khmer Republic most of the other countries in the region have an average per capita national income which varies between US $ 160 to 500 with an approximate average of US $ 275. Since energy is an input to all productive operations there is a correlation with per capita national income i.e. the more the national income, the more energy is consumed per capita. Though this need not exactly be correct as the size and population vary in each country, the per capita consumption of energy is still one of the best available criteria for judging the degree of development of a country.

Due to lack of availability of data, China and a few of the other countries in the region have not been included in the study. Some of the countries are very highly populated as for example India with about 560 million people (1972).

Table II gives the population of the countries, the total electricity production, the total consumption of commercial sources of

Table II
POPULATION, ELECTRIC ENERGY PRODUCTION AND TOTAL ENERGY CONSUMPTION

S.No.	Country	Population ('000)	Density of population (per km²)	Total elect. production (10³ MWh.)	Total energy consumption (10⁶ tons coal equ.)	Per capita energy consumption (kg. coal equ.)
1.	India	563494	172	66385 (1971)	104.82	186
2.	Pakistan	56065	70	7449 (1971) P	9.66	158
3.	Burma	28885	43	654 P	1.66	58
4.	Sri Lanka	13033	199	995	1.91	146
5.	Bangladesh	53209*	–	920 P	2.40	32
6.	Indonesia	121630	82	2368 (1971) P	16.13	133
7.	Singapore	2147	3695	3144 P	1.90	885
8.	Malaysia	10910	33	3545 (1970)	5.66	491
9.	Philippines	39040	130	8666 (1970)	12.15	311
10.	Japan	106958	287	428577	344.55	3251
11.	Thailand	36286	71	5225 (1971)	11.05	305
12.	Laos	3106	13	64	0.24	79
13.	Rep. of Vietnam	15317*	–	1483	5.48	287
14.	Khmer Republic	5838*	–	1656 P	0.18	25

15.	Egypt	34839	35	8030	11.29	324
16.	Iran	30550	19	9100	29.14	954
17.	Turkey	•37000	47	11242	20.88	564
18.	Saudi Arabia	8199	4	1000 P	7.38	900
19.	Kuwait	914	51	3295 P	8.77	10441
20.	Iraq	10074	23	2261 (1971) P	6.46	642
21.	Jordan	2467	25	249	0.82	331
22.	Israel	3080	149	8478	8.35	2712
23.	Lebanon	2963	285	1545 P	2.64	889
24.	Syria	6673	36	1223	3.04	455
25.	Rep. of Korea	32360	329	12697	26.76	827
26.	World	3,782,000	28	5646,700	7408.68	1984
27.	Europe (excluding Eastern)	364,000	92	1295,832	1435.75	4000
28.	USA & Canada	230,690	11.9	2091,017	2659.80	11526

Source: U.N. Statistical Yearbook 1973

NOTE: 1. All the figures are for the year 1972 unless otherwise stated.
2. P: for sale purpose, does not include power generated by the industries for their own use.
3. * Figures are for 1963.

energy and the per capita energy consumption in kgs. coal equivalent.

Excepting Japan, Singapore and Middle Eastern countries the rest of the countries, with a total population of about 1000 million, have an average energy consumption of only about 200 kgs. coal equivalent. The population density per square kilometre varies in these from a low 4 to as high as 3700 but most of the countries are below 100. The per capita energy consumption of the world average of about two tons of coal equivalent is half that of Western Europe and about 1/6th of that of USA and Canada. The world average is about 10 times the Asian average (with the exception of Japan, Singapore and the Middle Eastern countries).

In order to assess the gap in the energy requirement and the need and scope for atomic energy, it is necessary to aim at a minimum requirement which it is desired to achieve. Since there is no known absolute criteria to guide us, we have to make certain assumptions in setting up this goal. One extreme way of looking at it is to hope for a target achievement of consumption in North America. Alternatively, one could set the Western Europe average or the world average as the goal. For the purpose of this paper, we may assume that the developing countries need a per capita energy consumption of about 4 tons of coal equivalent, i.e. of 32 MWh energy per capita per annum which was Western Europe's per capita average for 1972. The present level in Western Europe, is much above this figure already. (Even though the target of per capita energy requirement is put at 4 tons of coal equivalent, it is based on 1972 population figures. By the time actual increase in energy production takes place, the population would have increased considerably. For example, by the year 2000 it is expected that the population would almost be double the 1972 figure, which means that the per capita energy requirement of 4ßtons set now would effectively be only about 2 tons coal equivalent. This is the present world average.)

Table III gives data regarding known reserves of solid, fluid and hydro power available in Asian countries. The per capita energy consumption of 32 MWh per annum is very much higher than the hydropotential available which cannot contribute anything but a small fraction of the total energy requirements. The energy, therefore, has to come from coal or oil. Excepting Saudi Arabia, Kuwait, Iraq and Iran, in the countries for which information is available, to meet the minimum energy requirements of 32 MWh the reserves will last for only a limited number of years. For example, India would have exhausted its resources in approximately 50 years, Pakistan in about 11 years, Burma in one year, Indonesia in 8 years, Japan in 46 years and the rest of the countries in about 10 years or less. Kuwait has reserves to last, if consumed for their own needs, for about 4650 years, Saudi Arabia for about 920 years, Iraq about 185 years and Iran about 175 years. If the average of all the countries is taken, and if the reserves are used only between themselves, then it will not last for more than 50 years. But the above analysis is misleading, in the sense that the Middle Eastern countries as major oil producing countries are meeting the demand of most of the countries in the rest of the world and it is practically impossible for the Asian countries to spend their resources only among themselves. If one takes solid fuels only, the picture is still worse. The resources in the region are expected to last less than 30 years. This also does not give the real picture as the solid reserves available are located only in five or six countries and there is a minimum production capacity possible as will be shown later. Also, the question of transport and intercountry co-operation is involved.

It can be concluded that for many Asian countries, there are not adequate resources available of their own to produce the minimum per capita energy requirement to sustain a minimum essential quality of life without the massive infusion of energy from outside sources. Even then, the extractable resources available at reasonable costs to satisfy the energy demands seem to be only for a limited number of years.

The position could be analyzed from a different angle. Table IV shows the coal and lignite reserves and annual production of

Table III

ABSOLUTE RESERVES OF COAL, LIGNITE, PETROLEUM, NATURAL GAS, HYDRO-POWER AND PER CAPITA RESERVES

S.No.	Country	Population (million)	Total Solid Fuels (10^6 MWh.)	Total Fluid Fuels (10^6 MWh.)	Total Solid & Fluid Fuels (10^6 MWh.)	Per capita (MWh.)	Water Power 10^6 MWh/year	Per capita/year (MWh.)
1.	India	563.5	855237	1546*	856783	1520	320	0.56
2.	Pakistan	56.1	13988	5858	19846	354	***	***
3.	Burma	28.9	830	92	922	32	–	–
4.	Sri Lanka	13.0	–	–	–	–	11	0.84
5.	Bangladesh	53.2	–	–	–	–	***	***
6.	Indonesia	121.6	11760	19245	31005	255	0.81 (1966)	0.01
7.	Singapore	2.1	–	–	–	–	–	–
8.	Malaysia	10.9	–	2352	2352	216	5.36 (1967)	0.53
9.	Philippines	39.0	220	–	220	6	15.1 (1967)	0.38
10.	Japan	107.0	158396	121	158517	1481	131.9 (1967)	1.23
11.	Thailand	36.3	587	–	587	16	–	–
12.	Laos	3.1	–	–	–	–	–	–
13.	Rep. of Vietnam	15.3	–	–	–	–	60.0**	1.00
14.	Khmer Republic	5.8	–	–	–	–	–	–

15.	Egypt	34.8	-	7517	7517	216	15.0 (1960)	0.43
16.	Iran	30.5	8000	162134	170134	5578	-	-
17.	Turkey	37.0	15382	335	15717	425	57 (1967)	1.54
18.	Saudi Arabia	8.2	-	240220	240220	29295	-	-
19.	Kuwait	0.9	-	133918	133918	148797	-	-
20.	Iraq	10.1	-	59196	59196	5861	-	-
21.	Jordan	2.5	-	-	-	-	-	-
22.	Israel	3.1	-	11	11	4	-	-
23.	Lebanon	3.0	-	-	-	-	-	-
24.	Syria	6.7	-	2184	2184	326	-	-
25.	Rep. of Korea	32.4	9492	-	9492	293	16.0+	0.5

Sources: *i) World Power Conference: Survey of Energy Resources 1968*
ii) U.N. Statistical Yearbook – 1973

All the figures are for the year 1972 unless otherwise stated

For the countries where no figures are given, either the hydro-resources are non-existing or very minor

* Present indications are that the actuals may be much higher

*** Only combined figures for Thailand, Laos, Rep. of Vietnam and Khmer Republic available.

*** Figures not available separately for Pakistan and Bangladesh

+ Estimated from information given in the Proceedings of the Conference on the "Peaceful Uses of Atomic Energy" – Vol. I, 1971.

Conversion factors:
1 ton coal	8,000 K.Wh.
1 ton lignite	2,500 K.Wh.
1 ton petroleum	12,000 K.Wh.
1 m³ natural gas	10.6 K.Wh.

Table IV
COAL AND LIGNITE RESERVES AND ANNUAL PRODUCTION FOR MAJOR COAL PRODUCING COUNTRIES

S.No.	Country	Population (million)	Coal & Lignite Reserves (10^6 T.C.E.)	Annual Production Total (10^6 T.C.E.)	Annual Production Per capita (Kg.C.E.)	Per capita Energy Consumption (Kg.C.E.)	% Contribution of coal production to energy consumption
1.	U.S.S.R.	247.4	4,561,097	498.7	2015	4767	42.2 E
2.	China	800.7	1,011,220	400.0	500	–	– –
3.	U.S.A.	208.6	1,226,875	537.0	2571	11611	22.1 E
4.	India	563.5	106,900	75.7	134	186	72.0 E
5.	W. Germany	61.6	89,375	137.2	2227	5396	41.2 E
6.	Poland	33.1	50,385	162.6	4912	4556	107.8 E
7.	Australia	12.9	45,875	56.3	4364	5701	76.5 E
8.	U.K.	55.8	15,500	119.5	2141	5398	39.6 E
9.	Japan	107.0	19,790	28.0	261	3251	8.0 I
10.	France	51.7	2,810	30.6	591	4928	12.0 I
11.	Belgium	9.7	1,796	10.5	1082	6468	16.7 I

Source: UN Statistical Yearbook 1973.

T.C.E.: Tons Coal Equivalent
Kg. C.E.: Kilogram Coal Equivalent
E indicates exporting countries
I indicates importing countries

* Based on the assumption that the coal production is used entirely for internal consumption and there is no export

11 major coal producing countries in the world which altogether possess more than 90% of the world's known reserves. It can be seen from the last column of this table that even if we are to assume that the entire coal production in any particular country is used for its internal consumption only, the percentage contribution of coal to total energy utilization is less than 75%. In many cases, it is less than 40%. The cases where the figures exceed more than 40% can be considered to be those which are exporting coal significantly.

Table V gives details of coal and lignite reserves, present annual production, and years of exhaustion if coal and lignite contributes 25% to 100% of the total energy requirements. It is seen that only India and Japan have any significant deposits available and even their deposits are limited. For example, India's resources will run out in less than 200 years even if it meets only 25% of energy requirements. At present, the coal consumption pattern is somewhat difficult to correlate with the production of coal as many of the coal producing countries export coal to other countries. However, if it is assumed that 50% of the total energy consumption aimed at (i.e. 4 TCE) is contributed by coal, then in India's case the reserves will last less than 100 years.

Table VI gives the per capita production of energy from a possible exhaustion of the coal reserves in 100 years. To this is added the per capita production of energy from a possible exhaustion of the fluid fuel reserves in the same period. The per capita reserves of hydro potential also is given in the table and in the last column, the total per capita available energy per annum. It can be readily seen that only Iran, Saudi Arabia, Kuwait and Iraq exceed the projected minimum energy requirement of 32 MWh per capita per annum. In all the rest of the countries the deficit varies from 16 MWh to very high values. Even assuming there are certain errors in the assumptions and calculations, the wide gap existing between energy demands and availability in the South East Asian countries is very striking. The production in Iran, Iraq, Saudi Arabia and Kuwait for 1972 has been 2976, 3420, 1812 and 852 million

Table V

COAL AND LIGNITE RESERVES AND ANNUAL PRODUCTION

S.No.	Country	Coal & Lignite Reserves (10⁶ T.C.E.)	Annual production* (10⁶ T.C.E.)	Years of exhaustion if coal's share in per capita energy requirement is				Export or import
				1 T.C.E.	2 T.C.E.	3 T.C.E.	4 T.C.E.	
1.	India	106905	75.90	190	95	63	48	E
2.	Pakistan	1748	1.25	31	16.5	10	8	I
3.	Burma	104	0.016	3.6	1.8	1.2	0.9	I
4.	Sri Lanka	–	–	–	–	–	–	–
5.	Bangladesh	–	–	–	–	–	–	–
6.	Indonesia	1470	0.18	12	6	4	3	–
7.	Singapore	–	–	–	–	–	–	–
8.	Malaysia	–	–	–	–	–	–	–
9.	Philippines	27	0.04	0.7	0.35	0.2	0.2	–
10.	Japan	19790	28.02	185	92.5	62	46	I
11.	Thailand	73	0.10	2	1	0.8	0.5	–
12.	Laos	–	–	–	–	–	–	–
13.	Rep. of Vietnam	–	–	–	–	–	–	–
14.	Khmer Republic	–	–	–	–	–	–	–

15. Egypt	–	–	–	–	–	–	–
16. Iran	1000	1	33	16.5	11	8	–
17. Turkey	1923	6.25	52	26	17	13	–
18. Saudi Arabia	–	–	–	–	–	–	–
19. Kuwait	–	–	–	–	–	–	–
20. Iraq	–	–	–	–	–	–	–
21. Jordan	–	–	–	–	–	–	–
22. Israel	–	–	–	–	–	–	–
23. Lebanon	–	–	–	–	–	–	–
24. Syria	–	–	–	–	–	–	–
25. Rep. of Korea	1187	12.4	37	18.5	12	9	1

Source: U.N. Statistical Yearbook 1973

Table VI
PER CAPITA ENERGY AVAILABLE FROM VARIOUS RESOURCES

| S.No. Country | Population (million) | EXHAUSTION PERIOD OF 100 YEARS | | | | Per capita reserves of hydro power (MWh.) | Total per capita energy available per annum (MWh.) |
| | | Coal & Lignite | | Fluid Fuels | | | |
		Annual production (10^6 MWh.)	Per capita production (MWh.)	Annual production (10^6 MWh.)	Per capita production (MWh.)		
1. India	563.5	8552	15.17	15.46	0.03	0.56	15.76
2. Pakistan	56.1	140	2	58.58	1.04	***	3.53**
3. Burma	28.9	8.3	0.29	0.92	0.03	–	0.32
4. Sri Lanka	13.0	–	–	–	–	0.84	0.84
5. Bangladesh	53.2	–	–	–	–	***	–
6. Indonesia	121.6	117.6	0.97	192.45	1.58	0.01	2.56
7. Singapore	2.1	–	–	–	–	–	–
8. Malaysia	10.9	–	–	23.52	2.15	0.53	2.68
9. Philippines	39.0	2.2	0.06	–	–	0.38	0.44
10. Japan	107.0	1584	14.80	1.21	0.01	1.23	16.04
11. Thailand	36.3	5.87	0.16	–	–		1.16+
12. Laos	3.1	–	–	–	–	1.0+	1.00+
13. Rep. of Vietnam	15.3*	–	–	–	–		1.00+
14. Khmer Republic	5.8*	–	–	–	–		1.00+

15. Egypt	34.8	–		75.17	2.16	0.43	2.59
16. Iran	30.5	80	2.62	1621.34	53.15	–	55.77
17. Turkey	37.0	154	4.16	3.35	0.09	1.54	5.79
18. Saudi Arabia	8.2	–		2402.20	292.95	–	292.95
19. Kuwait	0.9	–		1339.18	1487.97	–	1487.97
20. Iraq	10.1	–		591.96	58.60	–	58.60
21. Jordan	2.5	–		–	–	–	–
22. Israel	3.1	–		0.11	0.04	–	0.04
23. Lebanon	3.0	–		–	–	–	–
24. Syria	6.7	–		21.84	3.25	–	3.25
25. Rep. of Korea	32.4	95	2.93	–	–	0.5	3.43

* Figures are for the year 1963

** Excluding hydro power

*** Figures not available separately for Pakistan and Bangladesh

+ Based on average figure for hydro power for Thailand, Laos, Rep. of Vietnam and Khmer Republic.

MWh[1] respectively which means that these countries would stand to deplete their oil reserves in about 55 years for Iran, 70 years each for Saudi Arabia, Kuwait and Iraq. There is also a great demand now for coal to be exported from coal producing countries to other countries in view of the energy crisis. This will, to a certain extent, reduce coal availability in the coal producing countries.

Table VII gives the present installed capacity per thousand population and the electrical production per capita in the countries under consideration. It essentially shows how little installed capacity the Asian countries have (excepting Japan, Singapore, Kuwait and Israel). Per capita installed capacity varies from a very low .007 to about .14 KW with an average in the range of .04 KW. This can be compared with the installed per capita of 2 KW in USA and 1.4 in UK (1972). Most of the countries have installed capacities less than 2000 MW and many among them especially the South East Asian countries have less than 1000 MW and a few less than even 500 MW. In these countries especially the South East Asian countries, such a situation poses problems if they try to introduce large blocks of power from a nuclear station or high capacity thermal station without adequate transmission and distribution systems.

How can this great gap in the Asian scene between needs, existing production and possible production of energy from their own reserves be overcome? Clearly, nuclear energy will have a role to play in the future. Each of the countries will have to decide on their policies and priorities. Japan with a per capita energy availability of only about 16 MWh on a 100 year exhaustion basis of its resources is already consuming about 26 MWh, which means that considerable energy imports are already taking place. It is an example of a country with not enough reserves of energy resources, meeting its energy demands from alternate sources by import. Likewise, each country has to make a cost benefit analysis and develop its own strategies based on financial resources and other factors. Such studies must take into consideration the possibility of tapping the maximum from the 'income' sources of energy.

Since, for obvious reasons, the situation with respect to Japan and Middle East is different, it might be more relevant to analyze the requirements of the rest of the countries in the region. While it is clearly beyond the scope of this paper to discuss in detail all the possibilities, an overall view of the situation can be projected.

Before taking up these countries individually, however, it is of interest to have a look at the uranium resources in the region. India, Japan and Turkey have identified uranium resources with India having the maximum. But even India's reserves of uranium are only minimal. It is quite possible that some of the countries like Malaysia, Indonesia, Burma, Thailand etc. have not been surveyed in detail for their uranium resources. Therefore, the general assessments made now may need correction in the future.

In the middle 60s' the position of uranium resources were considered in detail and the available resources were assessed at prices up to $ 66 per kg. of U_3O_8. Subsequently, assessments were made up to $ 220 per kg. of U_3O_8 and in some cases even up to $ 1100 per kg[2]. These assumptions have been made to indicate that nuclear reactors, being highly capital intensive, the sensitivity of fuel costs even up to high cost of uranium is not critical when compared with power producing plants based on fossil fuels. It has been calculated that in the case of light water reactors the bus bar costs (of electricity) would change only by about 1 mil. per KWh if the price of uranium is increased from $ 22 per kg. to $ 66. The increase would change only by 5 mil. if the price is stretched up to $ 220 per Kg. of U_3O_8. In the case of breeder reactors the price increase of uranium is comparatively little felt in the bus bar costs.

Table VIII gives a rough calculation of the maximum energy available from world's uranium reserves based on 30-100 years of exhaustion of resources. If we take the world as a whole, at a cost of uranium at less than $ 33 per kg. of U_3O_8, per capita energy equivalent in MWh per year for even a 30 year exhaustion will be less than 2.5 MWh. However, there is a quantitative increase if the same is used in breeder reactors.

The above information is only of theoretical interest to many

Table VII
ENERGY RESERVES, INSTALLED CAPACITY AND ELECTRICAL PRODUCTION

S.No.	Country	Population (million)	Area (sq.km)	Energy reserves Solids & fluids Total (10⁶ MWh.)	Energy reserves per capita (MWh.)	Water power potential Total (10⁶ MWh.)	Water power potential Per capita (MWh.)	Installed capacity Thermal+ (MW)	Installed capacity Hydro (MW)	Installed capacity Total (MW)	KW per 1000 population	Electrical production per capita (KWh.)
1.	India	563.5	3280483	856783	1520	320	0.6	11285	6613	17900	31.8	117.8
2.	Pakistan	56.1	803943	19846	354	***	***	1264	586	1850 (1970) P	32.9	132.7
3.	Burma	28.9	678033	922	32	–	–	155	103	258 (1971)	8.9	22.6
4.	Sri Lanka	13.0	65610	–	–	11	0.84	86	195	281	21.6	76.5
5.	Bangladesh	53.2	142776	–	–	–	***	–	–	547 P	10.3	19.3
6.	Indonesia	121.6	1491564	31005	255	0.81	1.01	605	315	920 (1971)	7.6	19.4
7.	Singapore	2.1	581	–	–	–	–	727	–	727 P	346.0	1497
8.	Malaysia	10.9	329749	2352	216	5.36	0.53	728	293	1021 (1971)	93.6	325
9.	Philippines	39.0	300000	220	6	15.10	0.38	1627	549	2176 (1970)	55.8	222
10.	Japan	107.0	372154	158517	1481	131.90	1.23	64562	20734	85296	797.0	4005
11.	Thailand	36.3	514000	587	16	–	–	930	475	1405 (1971)	38.7	144
12.	Laos	3.1	236800	–	–	–	–	14	33	47 P	15.2	20.6
13.	Rep. of Vietnam	15.3*	173809	–	–	60.0**	1.0	–	–	838 P	54.7	
14.	Khmer Republic	5.8*	181055	–	–	–	–	70	–	70 (1971)	12.1	

15.	Egypt	34.8	1001449	7517	216	15	0.43	–	–	4004	115.0	230.7
16.	Iran	30.5	1648000	170134	5578	–	–	2007	800	2807 (1971)	92.0	298
17.	Turkey	37.0	780576	15717	425	57	1.54	1865	877	2742	74.1	303.8
18.	Saudi Arabia	8.2	2149690	240220	29295	–	–	325	–	325 (1971)	39.6	122
19.	Kuwait	0.9	17818	133918	148797	–	–	700	–	700 (1971)	777.8	3661
20.	Iraq	10.1	434924	59196	5861	–	–	840	–	840 (1971)	83.1	223.8
21.	Jordan	2.5	97740	–	–	–	–	50	–	50 (1971)	20.0	99.6
22.	Israel	3.1	20700	11	4	–	–	1470	–	1470 (1971)	474.2	2734
23.	Lebanon	3.0	10400	–	–	–	–	175	246	421 (1971)	140.3	515
24.	Syria	6.7	185408	2184	326	–	–	–	–	354	52.8	182.5
25.	Rep. of Korea	32.4	98484	9492	293	16	0.5	2568	340	2908 (1971)	89.7	391.8

Source: i) *U.N. Statistical Yearbook 1973*
ii) *U.N. Statistical Papers Series N J 16 "World Energy Supplies 1968-1971"*.

*Figures are for 1963 only
**Average figure for Thailand, Laos, Rep. of Vietnam, Khmer Republic
***Separate figures for Pakistan & Bangladesh not available
+Includes nuclear installed capacity also in countries having nuclear power plants

NOTE: 1. Figures are for 1972 except where mentioned otherwise.
2. P indicates power for public use only i.e. does not include
power generated by industries for their own consumption.

Table VIII
ENERGY AVAILABLE FROM URANIUM RESERVES

S.No.	Type of reserves	Reserves		Possible annual generation to exhaust resources in			Per capita availability MWh./year for exhaustion in		
		$(10^9$ T.C.E.)	$(10^9$ MWh.*)	30 years $(10^8$ MWh.)	50 years $(10^8$ MWh.)	100 years $(10^8$ MWh)	30 years	50 years	100 years
1.	If used in light water reactors (Recovery cost $ 15/lb U_3O_8)	60-90	163-245	54-82	33-49	16.3-24.5	1.46-2.21	0.89-1.32	0.44-0.66
2.	If used in breeders in the same ore cost range	5000-7500	13600-20400	4530-6800	2720-4080	1360-2040	122.4-183.8	73.5-110.3	36.7-55.1
3.	If used in breeders with ore recovery costs up to $ 100/lb U_3O_8	50000-75000	136000-204000	45300-68000	27200-40800	13600-20400	1224-1838	735-1103	367-551

Source: International Atomic Energy Agency Bulletin Vol. 15, No. 5, 1973

* In terms of electric energy 1 T.C.E. = 2.72 MWh. of nuclear electrical energy.
** Based on world population of 3700 million as in the year 1972.

developing countries, most of them not having any uranium and hence are unable to utilize it to the extent they desire. Also, most of these resources would go to feed the power requirements of the advanced countries.

The Indian situation is outlined briefly to give an idea as to how a developing country attempts to solve its vast demands for energy to the extent it can.

As a rule of thumb, it is assumed that 25% of the primary energy is used for domestic and commercial needs, 25% for industry, 25% for transportation and 25% for generation of electricity (gross). Based on a minimum energy demand of 32 MWh per capita per annum, in the context of our discussion, it can be assessed that 8 MWh are required for generation of electricity. This has to be contributed by all sources of electrical production including nuclear energy.

In India, the hydro potential for power generation is estimated at 41000 MW of which 6400 MW were operating in 1970. These sources are located remotely from the industrial areas and many of the rivers are monsoon dependent. Even assuming that the full hydro resources are tapped and used in the most efficient manner, still the maximum power availability from these sources is only about 560 KWh per capita per annum. This figure though higher than the present per capita production from all sources is very much lower than the projected demands. The coal and lignite reserves are assessed at 107 billion tons coal equivalent. It has been shown that the contribution from these sources is only about 15 MWh assuming 100 years as the basis of exhaustion. Of this only 50% is expected to go into electricity power generation; i.e. equivalent of only 7.5 MWh. But this assumes a production of 1000 million tons per annum which is more than 10 times the existing production. A realistic basis would be to assume a maximum coal production tons of about 200 million tons per annum, which means that the contribution to the electrical demand could be about 1.5 MWh thermal only. We have already seen that the power production potential from existing oil reserves is negligible

90

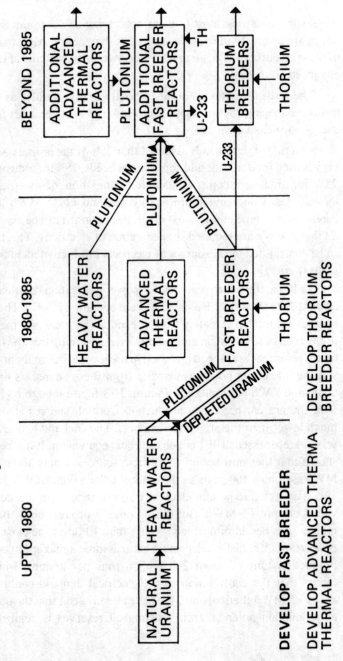

Fig 1. NUCLEAR POWER STRATEGY IN INDIA

though there are hopes of a much greater contribution from oil in the future. Thus the total maximum possible from all these sources seems to be less than 3 MWh. Even if we assume a 400 million tons annual coal production, it will still not exceed a total figure of 4 MWh from all sources. This leaves a deficit of 4-5 MWh to be filled by other sources of energy for electrical production. Certainly, nuclear energy is destined to play a major role in meeting this deficit. The magnitude of the problem comes to focus when we realize that to make up the 4 MWh per capita deficit (equivalent to about 1.2 MWh per capita actual electrical production) it is necessary to have an installed capacity of about 120000 MWe! This is quite a high capacity to be achieved. To add to this one has to remember that the population increase per annum is about 13-15 million. This highlights how great a problem the developing countries are being faced with in meeting even a tolerably basic energy demand.

Fig. 1 shows the nuclear power strategy in India. The thorium reserves of India are one of the largest in the world. It was realized from the very beginning that India's nuclear power programme will have to be ultimately based on these large reserves of thorium. The uranium available in the country is of low grade and is estimated to be sufficient to support a nuclear power programme of 500 MWe only for the life time of the reactors based on current reactor technology[3]. However, these uranium reserves, though small, are an important factor to trigger our fuel breeding cycle which alone can harness the vast thorium reserves. Taking these aspects into consideration, India's nuclear programme is planned in three phases. At present we are constructing and operating natural uranium-fuelled heavy water moderated and cooled reactors. One reactor is already in operation and five more are at various stages of construction in Rajasthan, Kalpakkam (near Madras City, South India) and Narora (Uttar Pradesh, North India). The natural uranium system was chosen, as in that case we would not need the sophisticated enrichment facilities. In fact the wisdom in India's choice of the heavy water system stands vindicated by developments in other countries.

The Indian nuclear programme envisages that the plutonium from the spent fuel of the thermal neutron reactors will be used to operate fast neutron breeder reactors which while producing power will also convert thorium into fissile U^{233}.

The technological development of fast neutron reactors has not been as fast as one would wish. This has been true the world over. However, it is expected that in about 10 years we shall be successfully operating our first fast breeder system.

In the third and ultimate stage of the nuclear power programme it is proposed to operate breeder reactors based on U^{233} which will convert more thorium into U^{233}.

It is very important to note that once the fast breeder programme is initiated the growth rate is governed by the plutonium inventory available at the start of the programme no matter how large the thorium resources are. This underscores the importance of our present thermal neutron reactor programme which will help build up a sufficiently large plutonium inventory.

Insofar as the other countries in the region are concerned, there is an urgent need (with the exceptions of the Middle Eastern countries) to explore and exploit new sources of energy. For most of these countries nuclear power may help in meeting their power needs. Conditions vary in each country. To start with, it is most important that a cadre of skilled scientific and technical personnel is created before embarking upon a nuclear programme. Industrial infrastructure has also got to be simultaneously developed. Nuclear technology should not be imported as a "black box" involving continual dependence on outside imported know-how, expertise and equipment. This is neither desirable nor viable from a long-term point of view. Concurrently, while looking for nuclear power development, the search for alternate resources has to be stepped up and existing resources exploited to the optimum. Regional co-operation is important in tapping available resources. For example, the Mekong river has great potential for hydro power. This can be exploited centrally or regionally, but if larger stations are set up the costs of generation and distribution would be lower. The countries of Indochina and Thailand, which in 1971-72 had only an installed capacity of about 2400 MWe with a projected demand of about 7000 MWe by 1985 and 22500 MWe by the year 2000[4], could very well consider the setting up of large capacity power plants and a common transmission system.

Table IX
POWER REQUIREMENTS
OF INDIVIDUAL COUNTRIES (MWe)

Country or area	Existing as of end 1968		Planned for implementation for years 1969-73		Additional requirements for years 1974-80	
	Hyero	Thermal	Hydro	Thermal	Hydro	Thermal
Malaysia*	264	274	Nil	420	252	270
Singapore	Nil	464	Nil	550	Nil	400
Sumatra	1.5	90.5	160	50	**	**
South Thailand*	Nil	70	Nil	60	27	30

* The details are for Western Malaysia and a small portion of Thailand in the south.

** There are preliminary suggestions that the Asahan hydro power project (Sumatra) with an installed capacity of 160 MWe (1974) be expanded by the additional 500 MWe; and a 200 MWe installed capacity thermal power station based on coal or lignite be set up near Palambang in Sumatra after 1974.

A study carried out in 1969 for Indonesia, Malaysia, Singapore and South Thailand shows that regional co-operation can be beneficially employed for power development[5]. Table IX gives the power requirements of the regions considered in the individual countries. It is clear that individually these countries would consider installation of single units of only about 100 MWe each. For unconnected development of the four national grids, nearly 700 MWe of generating capacity will have to be provided between 1974-80. If a single plant is considered, a 500 MWe capacity station (2x250 MWE) may be adequate, due to the economies possible for integrated operation of the four grids. The station whose output would be shared among the participating countries can be located at a site suitable and agreeable to these countries such that the distances for transmission to the countries are minimal. Alternatively, if a dual purpose plant is put up, in addition to power, the water demands for Singapore could also be met. Another possibility is for a central station located in Indonesia (Sumatra) not necessarily based on nuclear power but on fossil fuel. Detailed analysis of such possibilities are required before firm recommendations can be made.

Serious thinking has to be given also by the world community to explore the possibility of increasing the energy output in the countries without apparent substantial natural resources. Vast areas in these countries may still require exploration to find out what the true resource potential is.

Questions are being raised whether nuclear power really is the answer for meeting the short, medium and long-range needs of any country. Since the oil crisis in late 1973, a potential of about 35 billion barrels of oil has been found, according to one estimate[6]. The important point is that this is based on 18 months of exploration only.

Protagonists of nuclear power contend that nuclear energy is cheaper than that produced from oil at current prices and that it is not tied up with oil locales. Uranium being more widely found, is according to them a preferred raw material. Also, for the ultimate development of thermo-nuclear fusion, thermal reactor development is considered necessary.

Against this, there are questions about nuclear risks and waste management, about large inputs of capital, etc. Between now and early 1990's, it is expected that one and a half trillion dollars will be spent if all the present nuclear programmes of the developed countries are to be fulfilled. Nuclear stations are becoming increasingly costlier i.e. up to 5 times compared to those ordered 6 years ago. The cost of uranium has almost tripled during the last two years and though this accounts for a small fraction of the energy cost, the fear is that its availability will be a restrictive factor in the long run.

The fast breeder technology is of course projected as a hope for reducing the cost of electricity produced from nuclear sources. But large-scale commercial application is envisaged for the year 1985 and afterwards. It is not possible to predict what would be the cost of energy then. Developed countries have spent billions of dollars on fast breeder R & D, and still more will be required by the time commercial operations become feasible. Also in terms of safety and long term radioactive waste management, fast breeders have been under critical scrutiny.

The major hope lies in the development of thermo-nuclear fusion as early as possible, because of its immense potential and safety. The commercial exploitation of fusion seems possible only by the year 2000. If a strategy can be developed so that humanity's future needs for energy in both developed and developing nations can be met by fusion and other important sources, such as solar energy, then, there is some hope for the large under-privileged population living in the Asian region.

The complexity of the problem make any treatment inadequate. But the purpose of the paper would have been served if it helps to expose the magnitude of the energy and power needs of a region which accounts for a large percentage of the world's population.

[1] Source: *UN Statistical Yearbook – 1973.* Values include natural gasoline, crude petroleum and natural gas.

[2] Source: *IAEA Bulletin* Vol. 16, No. 1/2, 1974.

[3] Source: *Energy, Systems, Development and Policy:* 7th J.N.Tata Lecture by Dr. H.N. Sethna, Chairman, A.E.C. India, March 1974.

[4] Source: Committee for the Coordination of Investigations of the Lower Mekong Basin, Third Engineering Seminar, Vientiane, Laos, Nov. 69.

[5] Thomas, K.T., *Preliminary Study on the Feasibility of Regional Power Development including Desalination of Sea Water* – Report submitted as a special consultant to UN-ECAFE, Bangkok, February 1969.

[6] Source: *The Economist,* Volume 255, No. 6872, May 10, 1975.

ENERGY OPTIONS FOR LATIN AMERICA

JORGE SABATO *

I am not going to speak about energy options as in an energy conference, but about the rationality of introducing nuclear power to Latin America. Moreover, aside from the energy aspect there are and have been other reasons behind the decision taken by several countries in Latin America to go nuclear.

The Latin American countries can be grouped in three categories – in respect to their attitudes to and problems in developing nuclear energy. The first group consists of Argentina, Brazil and Mexico – in alphabetical order. The second group consists of Cuba, and some of the Caribbean islands, for geographical reasons, and the third group are countries of the so-called Andean Pact – Venezuela, Colombia, Ecuador, Peru, Bolivia and Chile, put together for institutional reasons. They are organized in a kind of Common Market in industry, commerce and many other things.

The countries of the first two groups – Argentina, Brazil and Mexico, and Cuba and the Caribbeans – are already engaged in nuclear development, while the countries of the third group – the Andean countries – are not yet engaged, and it does not seem as if they are going to be engaged by the middle of the 80's.

Let me start with a short description of the situation the countries of the first group – Argentina, Brazil and Mexico – find themselves in.

The situation in **Argentina** is as follows: at the present time one nuclear power station is operating producing 380 MWe per

---

* This is a shortened and edited version of the recording of Dr. Sabato's presentation to the Hearing Group.

year. It is a natural uranium heavy water reactor, but not a Canadian type. Another nuclear power-station is under construction, a CANDU type of 600 MWe; the construction of a third and fourth station for 600 Megawatt has already been decided on, and they are in the first stage of planning.

By 1990 Argentina must have in operation 10,000 Megawatt nuclear power, and by the year 2,000 the amount of nuclear power must be about 30 to 35,000 Megawatt. Why this rapid development in Argentina? From the point of view of energy supply, it is very clear. Argentina has little coal – the amount is only 1/2 million tons per year, which offers no solution to our energy problem. Argentina is not self-sufficient in oil (only about 90%), and the oil resources of the country are very limited and too small to supply the amount of energy needed in the next decade.

Hydro-electricity was insignificant in the past, but is now about 12% of the total power of the country. But once you estimate the amount of energy needed in Argentina in the next 25 years, you realize that the hydro-electric power if 100% used will only satisfy 25% of the future needs by 1990. The situation of coal is not going to improve very much, and the situation of oil is going to be more or less the same, if not going down. Thus in 1967 the decision was taken to go nuclear; there was no other energy option available considering the needs of Argentina.

In **Brazil** the situation is the following: There is one nuclear power station being constructed in Brazil now. This is a Westinghouse-type reactor of about 600 Megawatt using enriched uranium. A second nuclear power station with two or perhaps three reactors has already been decided and is going to be built in the same place. By 1990 the needs of Brazil seem to be in the order of 15,000 Megawatt nuclear. This again results from an analysis of coal, oil and iron supplies of the country. Brazil has a very powerful programme in hydro, in fact the biggest hydro-electric power station in the world is going to be built at the Paraña River between Brazil and Paraguay.

Many other hydro-electric resources are being explored and developed in Brazil. The country has some coal which is used for iron and steel production and for electricity. The oil situation in Brazil is worse than in Argentina, and a high percentage of the oil consumed in Brazil is imported. This of course is one of the reasons why the nuclear programme gathered such a tremendous momentum in the last year. The oil crisis had a significant impact on Brazil's balance of payment, and as a consequence the Brazilian government took the decision a few months ago to go nuclear at an extraordinary pace. Just at this moment it is about to sign a contract for eight nuclear power stations using enriched uranium, to be installed in the next decade. The problem is going to be to install a facility for enriching uranium in Brazil. This is a rational consequence once you have such a big nuclear development programme.

Mexico has a situation intermediate between Argentina and Brazil. The existing oil resources are larger than in Brazil assuring an oil supply as good as in Argentina. The situation of coal is worse than in Brazil. Mexico has a nuclear power station under construction – a boiling water type, enriched uranium, 550 Megawatt. The decision for another unit is being taken these days, and it seems that by the year 1990 Mexico will have a 10,000 Megawatt nuclear and by the year 2000 the total nuclear power will be 50,000 Megawatt. Thus you see that by the year 2000 Brazil will have the highest installation of nuclear power, followed by Mexico, and then Argentina. In these three Latin American countries together, there will be nuclear power of the order of 150,000 Megawatt by the year 2000.

In the second group there are two countries, **Cuba** and **Jamaica** which, for geographical reasons, are in a similar situation: isolation and lack of natural resources for electricity – there is not enough hydro, no coal, no oil. For that reason a nuclear power station would be one of the best solutions for Jamaica. The decision has not yet been taken, but according to my information it is forthcoming. Cuba has already taken the decision about two months ago to have a Soviet pressurized water type nuclear power station by 1979/80 of about 440 MWe. The complete dependence of Cuba on oil from foreign sources had originally led already in the late 50's to a decision to go nuclear.

In the third group the situation is the following: **Colombia** is very rich in coal, and the existing supplies are adequate for the next few years. It is not very rich in oil but it has enough for the time being. **Venezuela** is one of the biggest exporters of oil, so there are no immediate problems. But there is the awareness in Venezuela that oil alone will not solve all the economic problems of the future. life. It is better to take some actions on production in time and to protect the available natural resources.

Venezuela is therefore analyzing its possibilities. The same is happening in **Ecuador** where a lot of oil has been found, and in **Peru. Bolivia** and particularly **Chile** are open questions. Chile has a lot of hydro-electric energy and practically all of it is already under utilization.

Let me add something about the other reasons which are behind the decisions of some of these countries to go nuclear. One of the very important driving forces which had led up to these decisions were questions like this: what had happened to oil and electric power production in our countries? The past experience was that oil and electricity were resources that were exploited by others. The sentiment today is that history is not going to be repeated. In respect to nuclear energy this means that we are not prepared any more to just wait and see what the powerful nations are going to do, to stand still until they decide to come to our country. One can imagine that at the beginning of the worldwide development of nuclear energy it was important to know whether there was uranium on one's territory before somebody was coming to look for this increasingly important natural energy resource.

Furthermore I would say that in the case of nuclear energy, it was a field that really inspired the idea of self-reliance in the technological sector that is now spreading all over the world. This idea involves some kind of capability to take your own technological decisions in this field. Of course you can commit many mistakes by taking your decisions in that way. But at least you are the owner of your mistakes and not suffering the mistakes of somebody else.

There was also the problem of nuclear weapons of course. This problem has never come into the picture very clearly because the countries which have nuclear weapons do not talk about them, they just build them, they experiment, and the other countries keep discussing them.

It is well known that two countries in Latin America did not sign the Non-Proliferation Treaty, namely Argentina and Brazil. Of course many people are very unhappy about this: we have not behaved well in the world community. In the case of Argentina, I know the rationality. For Argentina one attitude in foreign policy is not to discriminate against any other country. Concerning the Non-Proliferation Treaty, the position of Argentina is quite clearly presented to the United Nations. The treaty discriminates between those who have and those who do not have nuclear weapons. Those who already have nuclear weapons are free to keep developing them without any restriction but the others are prevented from getting them, and this is completely unfair from the point of view of energy politics.

It is not that we in Latin America are doing anything yet to produce weapons, it is just that we also wish to have our own capacity to judge, our own technical autonomous capability to decide what is best for our countries in that matter. For that reason we try to learn and develop as much as possible.

There is finally a third point that is having a bigger impact every day in Latin America. It is the fact that nuclear energy was introduced in our countries, at least in Argentina first and then in Brazil, according to a different pattern than it had been the case with electricity or oil technology. In the latter two cases the technologies were introduced about 50-60 years ago by buying a "black box". Nuclear energy follows a different approach, what we may call a "grey box". It is not a "white box", but it is not a "black box" either that you buy just blindly. Now you ask the proprietor to open the box and to understand what is inside the box. You try to learn and you try hard to develop your own local capability. This procedure was followed very much by the Indian-Canadian experience and we have been following a similar pattern in Latin America.

The fact that the first nuclear power station in Argentina, the first nuclear power station in Latin America, was not a "black", but a "grey" box has also tremendous consequences in other sectors of industry and of public works. Our countries are not accepting any more the importation of technology as it was in the past in the petro-chemical industry, in iron and steel manufacturing, in gas industry, and in many other activities.

IS NUCLEAR POWER COMPATIBLE WITH PEACE?

The Relation between Nuclear Energy and Nuclear Weapons

JAN PRAWITZ

An important issue in the current nuclear power debate is the fact that the plutonium produced together with energy in most nuclear reactors, might also be suitable as the explosive substance in atomic bombs. An extensive installation of nuclear power stations would over a long period of time result in the build-up of considerable stockpiles of surplus plutonium and thus produce the fundamental prerequisites for atomic bomb-making in a large number of countries of the world. In the view of many, such plutonium would sooner or later be used for bomb-making and subsequently for nuclear aggression. The necessity of prohibiting this possibility would, according to this view, greatly strengthen the conclusion that nuclear power must be stopped in all countries, thus removing the very basis for bomb-making.

If peace and nuclear power are incompatible, it is reasonable to assume that everybody will be for peace. But is that the pertinent problem? Is it possible to handle plutonium in such a way as to have both nuclear power and a peaceful world?

As the organization of the international society is primarily based on sovereign states, the analysis of the plutonium problem is better divided into two parts. One is the possible proliferation of nuclear weapons among national states. The other is the possibility that terrorists or sub-national groups might get access to fissile material and make their own nuclear explosive devices. It is also necessary to extend the problem to include the protection of highly enriched uranium-235, in addition to plutonium.

Technological background

When plutonium is produced along with power in nuclear reactors, it is fixed to highly radioactive fuel elements, fatally dangerous to handle except in special facilities. Before separation from the spent fuel, plutonium would thus watch itself. If after a cooling period used fuel is processed, plutonium may be produced in chemically pure form and could either be recycled as fuel in reactors or stockpiled. Only part of the plutonium produced in the future would thus be available in stockpiles. How big these stocks will grow will depend on a number of factors, but their content could very well reach a considerable size in the 1980's.

This plutonium would be reactor-grade, i.e. contain up to 30% concentration of the isotope plutonium-240. Weapon-grade plutonium should have no more than a few per cent of that isotope. The higher the concentration of Pu 240, the less predictable and weaker the explosion yield of the bomb. Weak in this context would mean weaker than most military nuclear weapons but large compared to most conventional charges.

Even if power reactors, optimally run, would produce plutonium unsuitable for bombs, it would be quite possible to run a reactor differently to get a "better" plutonium. In addition weapon-grade plutonium could be produced in research reactors and in special plutonium production reactors.

While plutonium is itself a nuclear fuel of great potential value all reactors are at present essentially fuelled with uranium, either natural or slightly enriched with the isotope uranium-235. This uranium is not usable for bomb-making – that would require almost pure Such weapon-grade uranium is now very scarcely available because it is used in a few research reactors only and because the facilities to produce it exist only in the nuclear weapon states. However, if uranium enrichment technology later becomes available to many nations a U 235 problem will be added to the Pu problem.

The terrorist case

For a terrorist or other sub-national group wanting their own atomic bomb, the most practical approach would be to steal bombs from a military depot. This would provide reliable bombs of high and predictable yields and avoid the cumbersome and difficult design and fabrication problems. If that is not possible the second best approach would be the seizure of weapon-grade uranium from a military stockpile, because uranium is much easier to handle than plutonium. If that also is impossible the third approach would be to steal weapon-grade plutonium from a military stockpile. As clandestine production of fissionable material in hidden facilities would be out of question for sub-national organizations, the last and least attractive possibility would be seizure of reactor-grade plutonium from civilian stocks. It is only that fourth possibility that is directly connected to the civilian nuclear power debate.

Should the terrorists choose the fourth possibility they will face the problems of designing and making the bombs themselves, which is not as easy as is usually claimed, particularly if the plutonium to be used is reactor-grade. It is true that most of the nuclear physics involved is available in the unclassified literature; but beside that there will always be a fair share of difficulties to overcome in the intricate and unromantic disciplines involved, i.e. electronics, chemistry, shaped charges, metallurgy etc. before the entire process would produce a nuclear bomb.

There are examples where students have designed an atomic bomb in the library of a university. But the designs we have seen have been very primitive and would probably not produce a bomb of nuclear size (or perhaps only a small one), should the student survive the manufacture! This is not a critical observation, however, because there is always the possibility that a group of qualified scientists or a professional bomb-maker who changed his loyalty will do that part of the job. Therefore, the solution to the plutonium problem must be physical protection of the fissionable material.

Fortunately, the prospects for solving this problem seem hopeful for two reasons. One is that it has been positively demonstrated for several decades that it is possible to protect effectively the huge arsenals of nuclear weapons and the military stocks of weapon-grade fissionable material. And there is no reason why the military protection procedures could not easily be applied also to civilian material. The second reason is that denying sub-nationals access to the bomb would be in the interest of all governments, without exception; accordingly, an international agreement on minimum standards for protection procedures is now being prepared.

It is sometimes claimed that guarding the plutonium-stocks of the future would require such an effort that countries would have to turn themselves into police-states. This is clearly exaggerated. Protecting nuclear weapons has been successfully performed in several nuclear weapon states for a long time without turning them into police-states, at least for that reason.

International Proliferation

While the sub-national proliferation problem seems to be manageable and can probably be successfully solved, the prospects for denying access of nuclear weapons to national states seem to be less predictable. A general and widespread application of nuclear energy would furnish a number of countries with a good part of the necessary physical basis for bomb-making. The Non-Proliferation Treaty (NPT) was intended to take care of this problem before the nuclear industry had turned the many have-nots into potential haves. The treaty prescribes that haves shall not transfer any nuclear weapons to any recipient whatsoever and that have-nots shall continue as have-nots. The treaty prescribes further that the International Atomic Energy Agency (IAEA) shall verify that the peaceful nuclear industry is not used for bomb-making by have-not parties. This safeguard system is now in operation. But the treaty being essentially an expression of political intent has a number of weaknesses. Firstly, the subscription to it is far from universal. About 90 states are parties but a number of countries having a significant nuclear industry remain outside.

Among these are haves like France, China and India and have-nots like Japan, Israel, South Africa, Brazil and Argentine, to mention a few.

Secondly, the safeguard system can only detect but not prevent the diversion of fissionable material to bombs. On the other hand the IAEA inspectors seem effective enough to be able to discover violations of the treaty at such an early stage that political counteractions can be properly organized, for instance by the UN Security Council. In addition, safeguards apply uniformly only in NPT-countries. They apply in part in a number of non-party states and not at all in some countries.

In addition, national states would of course have much better possibilities than sub-nationals, in both physical and economic terms, to carry through a nuclear weapon program. A state might for instance change the operation schedule of its nuclear power stations to produce a higher grade plutonium. And it might complement its nuclear fuel cycle to achieve more selfsufficiency in fissile material. But a state would on the other hand probably want their bombs for other purposes than would terrorists, and they would, therefore, need many more bombs and more predictable ones.

In coping with these problems, it is necessary to realize that the political will to stop proliferation is sometimes single-sided. Certainly most governments would agree that proliferation of atomic bombs among additional countries would be unfortunate. But when it comes to writing off their own nuclear option, a number of governments apparently hesitate.

Therefore, the problem should be approached in two ways. One is to convince states, which can make the bomb and want to make it, to change their minds by promoting the NPT and encouraging as many have-nots as possible to subscribe to it. The other is to organize joint policies influencing and restricting the nuclear industry in non-party, have-not countries.

The main responsibility for promoting the NPT would fall on the nuclear weapon powers. They alone can reduce the risk of nuclear aggression and they alone can remove certain discriminatory features of the NPT by agreeing among themselves on substantial measures of nuclear disarmament. Only such measures would in the long run convince the hesitating have-nots. A few agreements of this kind (SALT I) have been reached but most of them remain to be realized, and the rate of progress is not impressive. It is true that nine more have-nots, including several with a significant nuclear industry (Euratom) became parties to the NPT in connection with the recent Review Conference of the parties to the treaty. But one should have no illusion that universal pacification of the atom can be achieved soon by means of nuclear disarmament.

In addition to promoting the NPT there has always been an interest among states exporting fuel and equipment for nuclear power stations to see to it that their export does not support or become involved in any nuclear weapon program abroad. This aspect has been increasingly stressed after the Indian nuclear explosion last year. It is reflected in the NPT by a provision banning such export to non-parties unless the exported fuel or equipment is covered by IAEA safeguards. This provision in fact extends the application of safeguards substantially beyond the NPT-group of states.

Furthermore, leading export countries are now discussing joint export policies including a demand that IAEA safeguards shall apply not only on what is exported but on the entire nuclear industry as a condition for transfer of fuel, equipment and technological know-how. To be effective such policies must be uniformly subscribed to by the main exporting countries. Also recommended is the establishment of multinational nuclear fuel cycle centers, i.e. combining chemical reprocessing plants, fuel fabrication plants, waste management installations and long term spent fuel storage under multinational management minimizing the risk of both national and subnational diversion of plutonium.

Other measures have also been suggested such as: dispersed stockpiling of weapon-grade materials organized by the IAEA, in order to minimize the physical availability of such material in any single country; increased efforts to solve the remaining problems in recycling plutonium as reactor fuel, in order to encourage the consumption of existing plutonium as reactor fuel; international ownership, Euratom-style, of all fissionable material in all non-nuclear weapon states; and the general strengthening and improvement of the intensity of the IAEA safeguards system.

Concluding remarks

This analysis has so far avoided two drastic measures that could altogether solve the problem of civilian plutonium stockpiles. One is the cessation and banning of all nuclear power production in all countries, a measure that would finally and completely remove the whole problem of civilian stockpiles. The other is changing the role of the IAEA from international verification of NPT obligations to supranational management of all nuclear activities for peaceful purposes in all countries, a measure that would go very far in securing that atoms for peace would be atoms for peace for ever.

I do not intend to use this opportunity to plead for or against nuclear power in general. But it is necessary to state that neither of the two far reaching measures indicated above seems at present politically realistic on a global scale. No leader of any country has enough power to decide on measures like that. And even if such measures could be agreed upon, the huge military stocks of weapon-grade material would remain as would all the nuclear weapons. Therefore great efforts both to diminish the possibilities of diversion of peaceful atoms to bombs, and to accomplish nuclear disarmament are our only realistic choice, because we will have to face the possibility that the nuclear power industry will continue to grow in several countries for several years to come and that the nuclear arms race will not stop by itself.

It will thus be necessary to reach agreement, within the next couple of years, among the sovereign states of the international society on a number of measures. A few that are presently under consideration shall be listed here:

- joint and uniform export policies requiring effective pacification of the recipient's nuclear industry as a condition for transfer of fuel, equipment and technological know-how;
- strengthening and improvement of the IAEA safeguards system;
- increased efforts to solve the remaining problems for recycling of plutonium;
- protecting stockpiles of plutonium and other special fissionable materials from seizure by terrorists or other sub-nationals;
- dispersed stockpiling of weapon-grade fissionable material organized by the IAEA;
- establishment of multinational nuclear fuel cycle centers and further measures to increase the international management of the nuclear power industry;
- a number of substantial nuclear disarmament measures.

It is certainly impossible to promise that such measures will be agreet to or that they will be fully effective if agreed. But as long as a nuclear power industry exists in a number of countries, such efforts would be necessary to make nuclear power more compatible with peace. The ultimate success could not be based on physical measures alone. There needs to be the combined political will of all countries to stop proliferation and to accomplish nuclear disarmament. In the long run this will be the essential factor.

Coming back to my initial question wether it will be possible to combine nuclear power with a peaceful world, the answer cannot be a clear yes. The terrorist problem could certainly be solved. But it is impossible to say that the proliferation problem would also be solved. It might or it might not, depending on the unpredictable political will of so many countries.

SECTION 3: ALTERNATIVE ENERGY SOURCES

FISSION ENERGY AND OTHER SOURCES OF ENERGY

HANNES ALFVÉN

The problem of how to satisfy the avalanching demand for energy in the world is attracting rapidly increasing interest. The energy crisis in the United States – whether real or manipulated – has stimulated the already lively discussion. The problem has a scientific-technical aspect. What energy sources are available now and in the future, and what are the ecological consequences of their use? It has an economic aspect. What price do we have to pay for energy produced in different ways? Finally, it is associated with at least two important world policy problems, one concerning the international competition for energy sources and the other concerning the relations between atomic energy and atomic warfare. Of special importance is that the large scale deployment of fission reactors is creating an abundance of nuclear material in the world.

When the atom bombs exploded over Hiroshima and Nagasaki, many of the scientists who had taken part in the Manhattan Project became frightened by the result of their work. They tried to satisfy their conscience in two ways. Some claimed that the horror of the bomb should put an end to all wars – the same thought Alfred Nobel expressed when he invented dynamite. The Vietnam war has taught us that this was not true. Others claimed that the development of fission had given mankind the ideal source of energy – the fission reactor – which should be of such benefit that it would overshadow the curse of the bomb.

In the United States the development of the atomic energy reactor began under extremely good conditions. A competent team of scientists, trained by their work on the bomb, worked in excellent laboratories which were well financed by the U.S. government. Big industries established for bomb manufacturing switched, at least in part, to the development and manufacture of reactors, a profitable area because the U.S. government paid and took all the risks.

Development of the atomic energy reactor spread rapidly to other countries. This fast reaction was largely due however to military considerations. In some countries, namely, the Soviet Union, the United Kingdom, France and later China, atomic bombs were actually manufactured, whereas in other countries only atomic reactors for peaceful use were built. However, in several countries the motivation for building reactors was, at least initially, a desire to keep open an option for making atomic bombs sooner or later. For different reasons – technical, economical or political – no other country has yet made atomic bombs but some may "go nuclear" in the not too distant future. The strong international opinion against atomic bombs (the Non-Proliferation Treaty) and the activity of the International Atomic Energy Agency (EAEA) have been and still are a rather efficient brake on the spread of nuclear weapons.

Energy Policy Decisions

Atomic energy thus received a flying start and development proceeded. A strong motivation for further investment in nuclear reactors was that the knowledge and technology already built-up must be utilized. The result was that in many countries energy policy decisions were distorted because the primary goal was not how to cover the energy need of the country but how to find an application for atomic energy.

From what has been said, it is clear that the nuclear industry partially received its internationally powerful position due to its association with the atomic bomb. In fact, the nuclear industry has been and probably still is supported by military subventions. The enriched uranium on which the reactors in most countries depend is a by-product of atomic bomb production, and a great deal of the development work for its manufacturing has been charged to military accounts. It would be interesting to obtain an objective clarification of whether atomic energy can be considered a cheap and competitive energy source without the more or less hidden subventions.

During the period of the development of nuclear reactors some opposition to this technology appeared; but from a technical, scientific point of view these objections seemed to be unwarranted. Certainly there were some unsolved problems, but these did not appear very serious. The prospects for atomic energy looked very promising. (As a personal declaration, up to a few years ago I was convinced that fission energy was the solution to the energy problem until fusion energy was ready.)

Objections to Fission Technology

The optimistic period for fission technology ended around 1970. There were several reasons for this.

1. It became increasingly obvious that plutonium and several of the waste products, especially radioactive strontium, are perhaps the most poisonous elements we know. For example, if introduced into the human body some of them tend to be deposited in the skeleton which they irradiate for a long time, increasing the risks of cancer even if the total quantity is only a small fraction of a milligram.

2. There are in nature a number of complicated biologic processes which enrich some of the radioactive waste products by a factor of 1,000 or 100,000. It is possible that even more efficient concentrating processes exist. Therefore, it is dangerous to deposit

radioactive waste anywhere in the biosphere, even if highly diluted.

3. Development of the breeder reactor was proceeding and the uranium reactors already in use began to be considered as a transition to the breeder. The breeder technology, which at least at present is mainly based on the uranium-plutonium cycle, means an enormous increase in the production of plutonium.

4. Hitherto the discussions of the waste problem referred to one or a few reactors. Now plans were made to use atomic energy to satisfy a substantial part of the world's energy needs, and that meant necessarily the mass production of radioactive waste and plutonium.

5. The waste products from one or a few reactors can be taken care of; but when a huge amount of such products is accumulated, a very serious problem appears because these products cannot be distroyed by any applicable technology. This has been demonstrated, for example, by the project to make a "nuclear repository" in a salt mine in the United States in Kansas, which had to be stopped because of leakage risks. Supported by competent geologists, the state of Kansas refused to accept the project.

At present there does not seem to be any existing, realistic project on how to deposit radioactive waste; but there are a multitude of optimistic speculations on how to do so. The problem is how to keep radioactive waste in storage until it decays after hundreds or thousands of years. The deposit must be absolutely reliable as the quantities of poison are tremendous. It is very difficult to satisfy these requirements for the simple reason that we have had no practical experience with such a long term project. Moreover, permanently guarded storage requires a society with unprecedented stability.

6. The ecological awakening has changed the basic view on technology. Hitherto a new technology has been allowed to take its short term advantages, leaving the long time disadvantages to

posterity. The essence of the ecological debate is that such a procedure cannot be allowed any longer. Applied to our case, the fission reactor produces both energy and radioactive waste: we want to use the energy now and leave the radioactive waste for our children and grandchildren to take care of. This is against the ecological imperative: Thou shalt not leave a polluted and poisoned world to future generations.

Fission Reactors and Ecology

Because burning coal and oil with present methods produces much air pollution, it has been claimed that fission energy is much cleaner, and hence from an ecological point of view is preferable. Some ecologists, however, call fission energy "the most dirty of all energy sources."

A single reactor or a few reactors which are carefully controlled are not likely to constitute a very serious ecological threat. Research reactors, either for scientific purposes or as technical "prototypes," are rather innocent. However, if nuclear technology spreads in such a way that a considerable fraction of the energy consumption of a country or of the whole world comes from nuclear reactors, the picture changes completely. The reason is that the production of nuclear energy is necessarily associated with the production of radioactive elements; and a very large production of nuclear energy necessarily means the mass production of radioactive poisons in quantities which are terrifying.

This is the basic reason for the opposition against the use of atomic energy, which has become a worldwide controversy.

On one hand, everybody must have the deepest admiration for all the ingenious precautions the reactor constructors have made in order to contain the radioactive products and to prevent them from reaching the biosphere. On the other hand, one must also respect the objectors, who are not driven by some "nuclear hysteria" but by a very well motivated fear of a new threat to the lives and health of their generation and future generations.

With reference to some arguments which have been stated or

published with regard to this controversy, it seems legitimate to state:

- It is not correct to claim that reactors are absolutely safe – because no technological product can ever be safe and no operator is absolutely reliable.

- It is not fair to claim that reactor accidents should be accepted in the same way as train and airplane accidents – because of the much more serious consequences which a reactor accident may lead to.

- It is not correct to claim that long time deposit of radioactive waste is not a serious problem – because this problem has not been solved as yet and, further, no one knows how to solve it on the required large scale if nuclear technology spreads and provides a considerable fraction of the energy consumed by a country or the world. On the other hand, one cannot exclude the possibility that future research may lead to acceptable solutions of these difficulties. There may be some chance that future discoveries may make fission energy acceptable; but we have not reached this state yet and no guarantee can be given that we will ever reach it.

Other Energy Sources

From what has been said, it seems obvious that with the present state of development fission energy should be accepted as a large scale energy source only if the need of energy is desperate and no other sources of energy exist.

How much energy *really* is needed in an acceptable society will not be discussed here. It is obvious that technological civilization is on its way to a limitation. We have reached a new stage in the development where our actions can no longer be dictated by a desire to increase the population or the consumption of a country. We shall confine ourselves here to the technical-scientific problem of how to solve the energy problem.

When comparing different sources of energy we must observe that a large amount of competent work and much money have been invested in fission energy, whereas very modest efforts have been made to develop other forms of energy. As stated above, the reason for this is because in many countries the problem has not been how to solve the energy problem, but how to give fission energy not only military but also civilian use. Accordingly, we must start thinking in a radically new way: *We must imagine how other sources of energy would have appeared today if research and development had been concentrated on them.* And what we could have expected of them in the future.

Apart from fission energy, the following energy sources are discussed as serious alternatives:*

Fossil fuels. It is often claimed that oil and natural gas sources will suffice only for the next 20 years; but in view of the fact that large new sources have been discovered, they may last much longer. In any case, there is coal sufficient for centuries. At present the environmental objections to fossil fuels are serious. However, one may hope that if research of the same quality and quantity that has been devoted to fission energy is directed to a non-polluting handling of fossil fuels, these sources may supply the world with energy for a long time and in a way that is tolerable for the environment. There are several new ideas of how to utilize the huge coal reserves in a clean way.

Fusion energy. A decisive difference between fission and fusion energy is that the fusion processes of interest result in non-radioactive end products. However, the intense neutron flux from a fusion reactor necessarily produces some radioactivity in the structure of the reactor. Also, a fusion reactor contains tritium, as an intermediate product which is radioactive, and this causes some leakage risk. There is no doubt that from an ecological point of view the fusion reactor is much less objectionable than the fission reactor. It is often claimed that technical fusion reactors will not be developed before year 2000 and, therefore, fusion energy

---

* Hydropower and wind and tidal energy are not considered here; hydropower is geographically limited, and wind and tidal energy do not seem to be available in large quantities as to be of importance except in special cases.

should not be mentioned in the present debate. The causality chain may be the reverse: as the breeder reactor lobby does not like the competition with the fusion alternative, this is eliminated by the claim that it belongs to a very distant future.

Solar energy. As each square kilometer of the Earth's surface receives as much energy from the Sun as a big fission reactor delivers (about one GW), we have here an inexhaustible and completely clean source of energy. It is, however, at present very expensive. New research results give reasons for optimism about the future economy of solar energy.

Geothermal energy. Geothermal energy means energy from the hot interior of the Earth. Outstreaming vapor and hot water in volcanic areas have long been used, for example, in Iceland, Italy and the Soviet Union. A new method, called the "hot rocks" method, has been suggested recently. Two holes close to each other are drilled until a hot region is reached, perhaps about 5 kilometers downward. The rock between the holes is cracked by some method; then water is poured down through one of the holes, and when it comes in contact with the hot rock it is vaporized and comes up through the other hole as steam. This may constitute an almost inexhaustible source of energy for all countries, especially for those countries in which the rocks have a high thermal gradient.

The Human Factor

Fusion, solar and geothermal energy sources are not yet sufficiently developed to ensure the solution of our energy problem; nor can we be sure that fossil fuels can be handled in a clean way as to satisfy the environmentalists. But, as both the Manhattan project and the Apollo program have shown, our science and technology are so powerful that if an intense effort is made, we can do almost anything we want in, say, 10 years – provided we are not in conflict with the laws of Nature!

It is obvious that the present international debate is creating new ideas; and it is very likely that energy sources will be found which may make fission energy unnecessary. It is, therefore, a mistake to concentrate energy policy on a line which initially seems attractive, but which in the near future may be considered obsolete and dangerous. As we have learned from history, it is a normal process that a technology which at a certain time seemed attractive is substituted with a better one at a later time.

As stated above, we cannot rule out that new discoveries will make fission technology acceptable by solving the safety and the waste disposal problems. However, as so much highly qualified work has already been devoted to these fields, this is not very likely. We now recognize that these problems are not the usual scientific-technological problems but are ones closely connected with the "human factor". To what extent can we trust that operators will really do what they are instructed to do? Do the social systems in different countries and in the whole world possess the unprecedented stability which the fission technology requires? This means that the basic questions fall outside the field of competence of the fission technologists.

Inseparable Twins?

Are the military and the peaceful atomic energy programs inseparable twins? Another strong objection to fission energy is derived from its close relation to atomic bombs. This relation is clarified in a number of reports of the IAEA and in the Stockholm International Peace Research Institute's *Yearbooks* as well as at a June 1973 SIPRI symposium on "Review of Nuclear Proliferation Problems". There is no doubt that the IAEA has had remarkable success in constructing a system of international inspection of nuclear material, according to the plans of the Non-Proliferation Treaty. However, not all countries have signed the NPT, and even for those countries which have signed it, IAEA authority is confined to inspection. There is no guarantee that efficient international sanctions would be applied to a country which breaks the Non-Proliferation Treaty.

According to existing plans, within 10 years the production of fissile material for peaceful purposes will be sufficient for the production of about 10,000 atomic bombs a year.

It has been claimed that the large plutonium quantities produced by the reactors cannot easily be used for manufacturing bombs because of the difference between "reactor plutonium", which is normally produced in the reactors, and "weapon-grade plutonium", which is used for atomic bombs. The former is obtained when a reactor is run in the economically most favorable way, with change of fuel elements about once every 18 months; the latter is obtained if the burn-out is limited to a few months. However, there is no serious technical difficulty for an establishment possessing complete fission energy equipment, to change to the production of weapon-grade plutonium. Moreover, even ordinary reactor plutonium can be used for making bombs, certainly somewhat clumsy and with less than maximum yield and less accuracy of performance, but anyhow terrible enough. Hence by stealing some 20 kilograms of ordinary reactor plutonium, a guerrilla or a criminal could obtain possession of the material for making atomic bombs.

It has also been claimed that the technology of manufacturing bombs is a very difficult process and requires knowledge of "atomic secrets". Today this is not true. Certainly, fabrication is not so easy that anyone could make an atomic bomb "in his garage" after having stolen a sufficient quantity of plutonium, as is said sometimes. But, according to what appear to be reliable reports, only a few engineers with a normal scientific and technical education and a good but not very fancy workshop are needed to make a bomb. This conclusion is quite reasonable from the point of view of the general experience of the technological development. What was a top scientific-technical achievement in 1945, when the first atomic bomb was made, may very well be an achievement that is rather easy to repeat in 1974. (However, the hydrogen bomb is still difficult to make!)

With all this in mind it is difficult to imagine how a proliferation of atomic bombs can be avoided in the future – when according to existing plans thousands or tens of thousands of atomic reactors will be working all over the world, and when the enormous production of plutonium from breeders has started. Indeed, it will be extremely difficult to prevent atomic bombs from falling into the hands of many groups of people who may like to use them for political or criminal purposes.

If we want to promote the spread of nuclear technology, the only way to avoid such a proliferation of nuclear bombs seems to be to impose very strict police control over the whole world. (Compare, Alvin Weinberg, "Social Institutions and Nuclear Energy". This will be difficult to achieve and does not lead us to a very attractive future society.

ENERGY OPTIONS FOR DEVELOPING COUNTRIES IN AFRICA

SHEM ARUNGU-OLENDE

INTRODUCTION

The development and use of energy is an integral part of the development process in any economy, whether it is an industrialized country or a developing country. Energy resource availability, supply and efficient use are therefore questions of particular importance to any country, subregion or region.

Energy consumption of the African countries has been growing steadily in the past years and a similar trend may obtain in the future. In most of these countries, non-commercial forms of energy, such as wood and wood products (e.g. charcoal), straw, baggase, animal waste still constitute the major source of energy, particularly in the rural areas. Of the commercial energy resources including coal, petroleum, natural gas and electricity, petroleum claims the largest share of total consumption (67.09 percent for the developing countries of Africa in 1972)[1]. It is, however, uncertain whether petroleum will continue to play the dominant role in the future in light of the recent sharp increase in its price.

To meet the expected increase in demand for energy, African countries will have to give serious thought to and adopt a number of options, both medium-term and long-term, that will best supply their respective requirements and that are best suited to their respective conditions. This is not going to be an easy task. Most of the African countries have as yet not undertaken a detailed inventory of their indigenous resources.

The need for an inventory and development of indigenous resources has been rendered the more urgent as a result of the above mentioned increase in the price of crude petroleum products, together with the general upward shift in the price of imported energy resources.

The rate of growth of installed generating plants in developing Africa has been of the same order of magnitude as that of total (commercial) energy in the 1961-1972 period. It is expected that the same general trend will obtain in the foreseeable future. Nearly fifty percent of this installed capacity is oil-fired; the recent increase in the price of crude petroleum products will, therefore, influence future electrical capacity growth in Africa, especially in respect of oil-fired thermo-plants.

In most of the African countries generating plants are, in general, still grossly underutilized, partly because (at least for some countries) generating plants are still isolated from one another and more substantially because of inherent problems in forward planning.

Although over 70% of Africa's total population lives in rural areas, the supply of energy in these areas is inadequate by any standards. Any appraisal of Africa's energy supply options must give serious attention to the energy requirements of these areas. The widespread use of wood and wood products in these areas results in deforestation, with adverse consequences on the soil, water table and general climatic characteristics of the affected areas.

COMMERCIAL ENERGY CONSUMPTION[2]

The combined consumption of commercial energy in the whole of Africa amounted to 134 mtce (million tons of coal equivalent), which represented only 20.4 percent consumption of the developing countries for the same year, but which nearly doubled the 70 mtce consumed in Africa in 1961.

There were considerable differences and disparities between the levels of energy consumption and rates of growth in the developing countries of Africa, on the one hand, and South Africa, on the other. For example, South Africa alone consumed 44 mtce and 65.33 mtce in 1961 and 1972 respectively, accounting for about half of the total consumption of Africa in each case. The average annual growth rate for the developing countries of Africa for the period 1961-1972 was 10.25 percent, a figure much higher than 3.65 percent for South Africa for the same period. There were also differences in the levels of energy consumption as well as growth rate between the developing countries of Africa.

Combined electrical energy consumption for Africa in 1972 was about 102.047 million KWh of which 59.081 million KWh was consumed in South Africa and only 42.966 million KWh in the developing countries of Africa. Electricity consumption rate of growth has lagged that of total energy in the 1961-72 period (8.7 percent per annum as compared to 10.25 percent per annum)[3].

While commercial energy consumption in developing Africa has followed the general average for all the developing countries, when viewed in a world perspective, developing Africa's share of world energy consumption has been a meagre 0.86 percent in 1961 declining to 0.58 percent in 1972.

ENERGY RESOURCES AND RESERVES

This section briefly discusses Africa's resources, reserves and production trends, where information is available.

Hydropower

While Africa has vast hydroelectric energy resources only a small fraction of this has been exploited to date. Its hydropower potential is of the order of 145,000 MW, which represents 26.16 percent of the world's total potential (of 555,000 MW). The sites are, however, not evenly distributed, with 56 percent located in Zaire, 8 percent in Madagascar and one percent in Zambia. The three countries alone thus claim as high as 65 percent of Africa's hydroelectric energy potential.

The installed hydropower capacity in Africa in 1972 was 7647 MW or 29 percent of the whole continent's installed electric power (hydro and thermal). This amounted to only 5.3 percent of the continent's hydropower potential. Great opportunities for the continent thus lie in the exploitation of the vast hydroelectric generating potential.

The continent produced, in 1972, 30.024 million MWh of hydroelectricity (a mere 2.16 percent of world production), which represented a contribution of 29.4 percent of the total of Africa's electricity consumption for that year.

The development of hydroelectric energy resources in Africa has been impeded by the comparatively large size of some of the sites and the heavy initial capital investment required by way of dams and generating facilities in relation to the size of the local markets. It is, however, hoped that in the future, with the electricity demand in Africa continuing at current levels, with petroleum (and petroleum products) prices soaring and with the expected sub-regional and regional cooperation in the development of hydroelectric energy resources, the pace of hydroelectric development and exploration in Africa will increase.

Large scale hydroelectric installations can be economically justified only where the site provides large hydraulic heat and larger flow rates. In addition to the larger waterfalls, there are a larger number of smaller ones in many countries in Africa. The development of small scale machinery that can economically be harnessed to supply local needs requires serious attention. Work has been carried out on the development of special types of hydrosystems that could be used for such purposes[4]. Further research and development work has been carried out by EDF (Electricité de France) which has reported encouraging advances in the field[5].

Hydrocarbons

(a) PETROLEUM

Significant petroleum development in Africa has taken place in the past two decades, exemplified by the discovery of large oil and gas fields in Algeria and Libya. The past few years have, however, witnessed decreased exploration activities resulting in a lower rate of discovery of crude petroleum than in the previous years.

For example, during the last three years, only 9.4 billion barrels of proved oil reserves were discovered in Africa as compared to 40 billion barrels discovered during the 1960's[6].

Substantial oil and gas reserves are to·be found in Algeria, Angola, Congo, Egypt, Gabon, Libya, Nigeria and Tunisia. Fairly good discoveries have also been made in Cameroon, Dahomey, Ghana, Senegal and Zaire, particularly offshore. In addition, recent exploratory results indicate that there are good prospects for discoveries in the coastal basins of Eastern Africa and Southern Africa. There are also good prospects of further oil discoveries in the basins with existing proved reserves such as the North African Atlas area, covering Tunisia, the North African Platform, stretching from Morocco to Egypt, and the Gulf of Suez and Red Sea basins.

As a result of the increase in the price of crude petroleum and products, it is expected that new areas in Africa will come under exploration. Increased exploration activity is in fact, already underway in many locations in the continent, either offshore or in the interior.

Africa's proved petroleum reserves are estimated at some 68.3 billion barrels which is 9.5 percent of the total world proved reserves (715.7 billion barrels)[7].

(b) NATURAL GAS

There are natural gas occurrences in Algeria, Congo, Egypt, Gabon, Libya, Morocco, Nigeria, Rwanda and Tunisia. Proved reserves of natural gas in Africa are estimated at 31,974 billion cubic feet, which represents some 12.3% of the world's proven reserves (of 2,555,064 billion cubic feet).

The discovery rate of natural gas has been much better than for crude petroleum during the last three years; about 122,000 billion cubic feet of proved natural gas reserves were discovered. The utilization of natural gas has, however, not caught on in Africa. Consequently up to 90 percent of produced natural gas is burned in the atmosphere.

The continent's crude petroleum production was in the 1950's about 100,000 barrels a day, with Egypt as the largest producer. In the early 1960's with Algeria and Libya joining the ranks, the continent's production passed the one million per day mark and continued to rise in the second half of the decade. By 1970 Africa's production averaged about six million barrels per day. In 1974, Africa's crude oil production was estimated at about 2 billion barrels, representing approximately 10.5 percent of the world's production for the same year. Nigeria became a producer towards the end of the 1960's and is today the single largest producer in the continent.

Although Africa as a whole is a net exporter of energy, the great majority of developing countries of Africa have remained net importers of commercial energy, principally petroleum. The burden and pressure on the balance of trade and balance of payments caused by the sharp increase in the price of petroleum and petroleum products has been excessive for many of these countries, and has in many cases resulted in unprecedented slackening of economic activity boardering on recession, in turn causing an aggravation of often acute unemployment problems.

Coal

Estimated reserves of coal in Africa are of the order of 16 billion metric tons (2.7 percent of world reserves of 590 billion metric tons) most of which is located in South Africa. It is extremely difficult to estimate the extent of coal resources in the developing countries of Africa, since exploration in these countries is far from complete. Significant reserves are also to be found in Algeria, Botswana, Morocco, Mozambique, Nigeria, Rhodesia, Zambia and Zaire, and to a smaller extent, Egypt, Madagascar and Tanzania.

One point is clear in relation to coal resources and reserves in Africa, and, for that matter, in any other part of the world: with the recent increase in the price of crude petroleum (and petroleum products) many hitherto marginal or sub-marginal coal deposits have now become viable in economic terms. This will definitely have a marked effect on the estimate of recoverable coal reserves.

In 1973 Africa produced 67.8 million metric tons of coal, equivalent to 3.9 percent of the total world production of 1.7 billion tons. Of the 67.8 million tons, 92 percent (62.4 million tons) and 4.4 percent (3 million tons) were produced in South Africa and Rhodesia respectively, the two countries together thus accounting for 96.4 percent of Africa's total coal production.

Oil Shale

Oil shale is a sedimentary, laminated rock containing kerogen, which contain up to 40% by weight of oil. Oil shale deposits have been reported in Madagascar (estimated reserves of oil: 4 million tons), South Africa (19 million tons of oil) and Zaire (15 billion tons of oil). Unestimated reserves are also to be found in Egypt, Gabon, Mali, Morocco, Niger, Somalia and Uganda.

It has hitherto been uneconomic to extract oil from shales in the face of low prices of conventional energy sources coupled with technical problems. These have combined to keep its development at low pace. The situation has changed drastically over the past 18 months, as a result of large increases in the price of crude petroleum and petroleum products. It is hoped that many African countries will give serious attention to the exploration and development of their national oil shale resources.

Tar Sands

These are sands impregnated with heavy oil that is too viscous to permit recovery by natural flowage into wells. In Africa tar sands are known to occur in Ghana, Ivory Coast, Madagascar and Nigeria.

The first large scale mining and extraction operation of tar sand deposits was in 1966 in Athabasca, Alberta, Canada[8]. With improvements in tar sand extraction technology now underway and with the high level of current prices for petroleum, it is apparent that the production of oil from tar sands has reached the point of economic viability. It is, therefore, likely that African countries having tar sand deposits will evaluate or reevaluate their tar sand deposits with a view to commercial utilization of the deposits.

Geothermal Energy

Geothermal energy resources are derived from the hot internal core of the earth. Natural manifestations include volcanoes, geysers, fumeroles, hot springs etc., while the sources harnessed by humanity include stream wells and hot water wells.

Successful geothermal production of electricity has been at Larderello (Italy), Wairakei (New Zealand) and the Geysers (USA), while the use of geothermal hot water is widespread in New Zealand and Iceland. In Africa geothermal exploration is underway in the Rift Valley in Kenya, Ethiopia and the territories of Afars and Usas (near Djibouti). Geothermal manifestations have also been observed in Uganda, Tanzania and Zaire. There is also a geothermal potential in Chad, Malawi and Somalia. Zaire has a plant, utilizing geothermal energy, in which low-pressure steam is used to drive a 220 KW generator set for mining purposes.

Although not widely distributed throughout the continent, geothermal resources are expected to provide a substantial portion of the energy requirements, at least in the countries that are endowed with this source.

The use of geothermal energy to produce electricity offers a number of attractive features; in comparison with the fossil-fuel fired plants, the requirement of a boiler is eliminated, thereby reducing both capital and maintenance cost for the equipment, and since no fuel is involved, pollution problems relating to combustion and fuel residues are eliminated.

On the other hand, geothermal fluids are, in general, corrosive and abrasive and thus require the use of special materials for handling. There may also be need for preventive maintenance procedures in the operation of the pipeline and power plant. All this can add appreciably to capital and operating costs. Furthermore, steam or hot water from a geothermal well cannot be conveniently stored as can, say, petroleum, nor is it easy to shut down a geothermal well. It is, however, possible to alleviate these problems through careful design and operating procedures.

Solar Energy

The whole of Africa possesses this form of energy in abundance and an increasing interest is now being shown in its development. The number of African states involved in solar energy research or development is increasing, though the activities undertaken so far are still in the experimental stage.

A solar energy research and development centre has been in operation for some time now at Niamey, Niger – the office is active in theoretical and experimental studies on concentrators, water pumps, and on the development and manufacturing of water bearers as well as in socio-economic studies in the application of solar energy in developing countries. In other African countries such as Chad, Mali, Senegal and Upper Volta work is at different stages on the design, improvement, installation and operation of solar collectors and pumps.

Solar thermal energy has been in use as a heat source from time immemorial, for drying crops and building materials, for baking and for food preparation by dehydration. The more recent, technologically advanced operations include solar heaters, solar stills, solar powered machinery and refrigeration devices; solar furnaces represent the farthest advance in the thermal utilization of solar energy, although the use of these had tended to be restricted to producing high temperatures for laboratory and experimental purposes.

Technical and economic factors have thus far limited widespread use of solar devices despite its advantages: inexhaustibility, enormous rate of supply to the earth's surface, universal distribution, neglegible cost (as raw material before conversion), and apparent freedom from environmental damage.

These technical and economic factors stem largely from the fact that solar energy is variable (geographic, seasonal, day-night and atmospheric interference) as well as of low intensity compared to that obtainable from such sources as fossil fuel etc. There is an urgent need for further research effort into the ways of overcoming the technical and economic hurdles in order to open the way for widespread practical utilization of solar energy, especially with regard to solar heating and cooling devices, solar pumps and devices for electricity generation.

In Africa, as in developing countries in general, there is need for the development of appropriate solar devices that are in harmony with both the cultural and physical characteristics of the rural (and sometimes urban) part of the country, based on the concept of user standards. In developing such devices it must be realized that once the appetite for energy has been aroused in these areas, demand tends to increase at a steady rate; the devices must therefore be based on concepts that the intended user understands and can expand as the need arises.

Biomass[9]

Photosynthesis converts electromagnetic energy into chemical energy. Plant life is based on this process, wherein, carbon dioxide and water are converted to carbohydrates and oxygen by utilizing electromagnetic energy absorbed in the pigments of the green cells of the plants. The chemical energy stored in the carbohydrates can be utilized to form more complex organic compounds by enzymic processes energized by the oxidation of a portion of the plant cell carbohydrates.

To date emphasis has been on photochemically produced energy only for food and cellulose production; this has led to the neglect of the important potential of plant life as a primary source of energy. In order to check this trend, it may be necessary to grow plants specifically for their fuel value. Algea would fit well into this latter category as they are among the most efficient in the photosynthesis conversion process; however, they require special resources for cultivation.

One of the most efficient of the higher plants, readily cultivated by common methods in many developing countries in Africa, is sugar cane, the juice of which can be extracted and fermented to produce alcohol; by itself or mixed with gasolene, alcohol could run internal combustion engines; even agricultural pumpsets could be run on alcohol, with only slight modifications to their engines. In the African countries with excess capacity of sugar cane production, its use as a potential energy source needs to be given serious consideration.

Nations with inadequate agricultural capacity or those which have difficulty in producing sufficient food can produce fuel from organic wastes through anaerobic decomposition. In anaerobic decomposition specific bacteria decompose carbohydrates to methane and carbon dioxide while releasing small quantities of heat. The residual organic sludge and carrier water retain the same organic nutrient and fertilizer as those resulting from aerobic decomposition.

Thus if the organic wastes can be gathered and introduced into anaerobic digesters without excessive energy expenditure a useful energy results. The resulting methane can be used to produce thermal energy for heating, cooking or powering heat engines; or it can be stored as preserved gas. It can also be catalytically converted to methyl alcohol, which is then readily stored as a liquid, to be used either immediately or at some later date.

In the energy-poor, food-poor regions of Africa, it may be necessary to practice the composing of organic wastes in order to produce fertilizers to aid food production. In the recent past artificial fertilizers, produced from hydrocarbons, particularly natural gas, of which African countries have a poor supply, have been at such low cost that compost fertilizers appeared uneconomic. With the current high energy prices, the situation may very well have changed in favour of composting. In reference to composting practices, anaerobic procedures should be introduced and encouraged, since not only is the nutrient value of the waste retained, but methane is also produced.

Anaerobic digestion for biogas and fertilizers production is already extensively practiced in certain areas of the developing world[10]. It is hoped that the methods will find widespread introduction in Africa, where rural areas would no doubt reap huge benefits from both fertilizers and methane.

Wind Energy

Wind energy, like solar energy, has been used by man from the dawn of history. Its use has been limited in comparison to other forms of energy due to its fitful availability, impossibility of transportation over long distances and inequality of distribution. Furthermore, because of the low density of the air in some areas, small capacity wind mills are bulky compared with other prime movers. In addition, the variability of wind speeds is not conducive to the desirable constance of voltage and frequency.

The problem of matching load to random availability of wind-power can be partially solved, especially in less isolated areas, by a parallel operation with a thermal (or hydroelectric) plant, which can then pick up or shed load so as to adjust to the difference between bad demand and the contribution from wind-power.

For isolated plants, the problem of load matching may be reduced by the introduction of some means of energy storage. One such scheme that offers promise is to use the variable power output wind to decompose water into hydrogen and oxygen both of which would be stored under pressure and recombined in a fuel cell to generate electricity on a steady basis. Further research effort is required to establish the best possible conditions for the proper functioning of such a system.

There are opportunities for utilizing energy in Africa, especially in the coastal areas where regular winds prevail.

Nuclear Energy

Insufficient exploration has been undertaken in Africa to allow for a good estimate of the uranium potential in the continent. However, a recent estimate by IAEA gives 390,000 short tons of reserves of U_3O_8 at a price lower than US10/lb (40% of the world's reserve of one million tons at the price range) and some 94,000 short tons at the price range between US 10/lb and US 15/lb U_3O_8 (12% of the world's reserve of 750,000 short tons at the price range[11].

In the developing countries a number of factors have precluded the widespread use of nuclear energy for electricity production. First, there is the technical problem of the limit on the generating plant in so far as it conflicts with the economics of scale.

The choice of maximum acceptable unit size is based on the permissible disturbance to a system which can be tolerated on the sudden loss of the unit, the disturbance being measured in terms of frequency deviation and whether or not automatic load shedding devices operate. In general the limit on the unit size is, on the average, of the order of 10% to 15% of the installed capacity or peak load.

This then limits the number of countries in Africa that can introduce nuclear generating units into their power system in the foreseeable future. In the IAEA Market Survey for Nuclear Power in Developing Countries (1973) it is reported that manufacturers indicated no interest in plant sizes below 400 MW-600 MW as this would be uneconomic. Taking into account the 10% to 15% limit referred to above this would require a system with an installed capacity into the range of 2667 MW to 6000 MW; few countries in Africa have power system installed capacity in this range for the foreseeable future.

The second factor hindering widespread nuclear energy use for electricity production is that few developing countries can affort to raise enough capital to cover the very high initial capital cost of nuclear power plants. The development of energy resources is capital intensive. Many of the developing countries do not possess adequate financial resources to meet the requirements for energy development and will, therefore, have to borrow from international markets or organization, or on a bilateral basis. It must be emphasized that these countries will be borrowing for not only energy resources but also other, even more pressing development requirements; so that, in effect energy may not receive enough funds for the development of the sector, and whatever goes to the sector will have to also meet the requirements other than electricity development in which nuclear energy may be part. Nuclear energy would therefore consume whatever little finds its way for electricity development.

Thirdly, in considering the introduction of nuclear plants in the developing countries of Africa, very serious thought must be given to the training of operating, maintenance and supervising personnel, and in the longer run, of the management and running of the plant installations. Lack of skilled personnel in these countries has been one of the bottlenecks for proper energy development even with respect to conventional sources of energy such as fossil-fuel fired plants.

The above misgivings apart, nuclear energy has promise in Africa in the longer run i.e. there is ample scope for the introduction of nuclear plants into a number of African countries' power systems. For one thing, the recent price increase in petroleum and petroleum products has dealt many of the African countries a stunning blow that has further aggravated their already adverse economic situation.

Moreover, in a sufficiently large number of African countries, the systems installed capacity shall have grown to large enough levels to take full advantage of economies of scale of the nuclear plants. Furthermore, it is expected that with the industrial and economic development of these countries, the high initial costs of the nuclear plants will be within the reach of some, which should have the capacity to train personnel for the operation and maintenance, supervision and eventual management of the installations and which should have the organizational capabilities required in the proper running of nuclear programmes. It is also hoped that advances in nuclear technology together with a concerted effort in the development of smaller scale units will bring nuclear plants within the reach of many developing African countries.

Some General Remarks

Although the price of crude petroleum and petroleum products rocketed some 18 months ago, there are so far no indications that the supply of petroleum and products of petroleum-deficit African countries will be threatened through shortages. There are in fact good prospects for further discoveries in the existing as well as new areas which together should be able to put more of this commodity at the disposal of the African countries.

It is, however, in the interest of individual African countries to undertake systematic survey and inventory of their energy resources in order to enable them to plan the most suitable resource development strategy for their needs. In this context special emphasis should be placed on the need for the formulation of effective energy policy at the national, subregional or regional levels; the policies should have clearly defined objectives and should be able to reflect the unique needs of each country, and at the same time be consistent with the regional and subregional set up.

It is also in the interest of individual African countries to formulate guidelines for, and implement, energy conservation. This could result in substantial savings in energy use for countries that are already hard pressed financially. Better conservation would thus stave off the currently observed gross under-utilization of electric generating plants in African countries (in the region of 3167 KWh/KW on the average for 1972) through better planning and maintenance procedures.

There is also a strong case for cooperation among African states in the search for new deposits of petroleum and natural gas; in the development, on a subregional basis, of hydro, coal, oil shale, tar sands and geothermal energy resources and in adopting measures (with oil producing countries) that would guarantee the supply of crude petroleum and products to them, under mutually acceptable conditions.

136

[1] World Energy Supplies (ST/STAT/SER.J/17).

[2] *The Role of Energy in the Development of Human Settlements in Africa,* by Economic Commission for Africa (E/CN.14/HUS/6), presented at Regional Conference on Human Settlements, Cairo, June 21-26, 1975.

[3] *ibid.*

[4] A.L. Johnson, *Non-Conventional Sources of Energy for Developing Countries,* Document prepared for the United Nations Committee on the Application of Science and Technology to Development (EAC.52/XX/CRP.7/ANNEX).

[5] Xavier Ract-Madoux, *Small Hydroelectric Plants Equipped with Encapsulated Generator Sets,* Paper submitted for the Interregional Seminar on Rural Electrification, New Delhi, December 2-12, 1971 (ESA/RT/Meeting IV/8).

[6] Economic Commission for Africa *Status of Energy Resources in Africa,* Draft Internal Report, May 1975.

[7] *ibid.*

[8] "World Energy Reserves, Supply and Demand" in *Projections of Natural Resources Reserves, Supply and Future Demand,* Document submitted to the Committee on Natural Resource, Third Session, New Delhi, India, February 1973 (E/C 7/40/Add.1).

[9] *ibid.*

[10] R.D. Laura et al, "Increased Production of Biogas from Cow Dung by Adding Other Agricultural Waste Materials", *Journal of the Science of Food and Agriculture,* Vol. 22 (April 1971); and
R.B. Singh, "Building a Biogas Plant", *Compost Science,* Vol. 13 No. 2 (1972).

[11] *ibid.*

SECTION 4: SOCIAL ETHICS OF NUCLEAR POWER

ETHICAL REFLECTIONS ON THE USE OF NUCLEAR ENERGY

ROGER L. SHINN

I. SKETCHES OF IMAGINARY FUTURES

Humanity is moving into a future so different from any familiar past, so loaded with portent and promise, so unpredictable that solid thinking about social and ethical decisions to be taken in that future is extremely difficult. For that reason an effort at loosening up imagination may be useful. Hence I shall sketch quickly four possible pictures of the world in the year 2100. Obviously all four cannot be true. I claim no truth for any one.

A. Technological Utopia

By the year 2100 the world has recovered from the near breakdown at the end of the 20th century. At that time the population of the globe had reached 6 billion, and about 10 million people starved in one year. Energy shortages disrupted the economies of several societies. There were sporadic revolutions in many parts of the world. Small countries and even terrorist gangs discovered that, even if they could not win a war, they could frighten the major powers into pay-offs.

In a sudden show of strength the U.S.A., the Soviet Union, and China collaborated to impose a world dictatorship. A combined military and scientific elite took over and imposed stern measures, which the world reluctantly accepted in order to avoid chaos.

Now, within slightly less than a century, the world population has been reduced to 5 billion people. Further reductions are planned, but it has been decided to move slowly because there is already a preponderance of older people. Governmental regulation of population is strict but fairly simple. The development of the 'ideal contraceptive' means that every youth is inoculated at puberty, and parenthood requires a specific decision and medical act.

The world is organized in regional economies. There is centralized planning, but regions are given some choices. For example, a region may choose to have low population with a high level of consumption or higher population with lower consumption. The inequalities of the past have been reduced but not ended.

There are no international armies, but a world police force and intelligence agency exercises strict surveillance. Activities destructive of the environment are prohibited. This means a close monitoring of inventions and industrial processes.

Historians now refer to everything prior to the 19th century as pre-history. The 19th and 20th centuries are called the time of primitive civilization, and they are divided into the coal age, the petroleum age, and the age of nuclear fission.

Since the second quarter of the 21st century oil reserves have remained "unused and unwanted in the ground as a consequence of a diminishing demand – arising from the abilities of man to harness and to use solar and fusion power and the preferences of consumers for these new forms of energy." (Peter R. Odell.) Large solar reflectors orbit the earth, remaining in fixed positions relative to the geography of the earth, and beam solar energy to receptors on the ground. Their use is precisely planned, not only to provide energy but to stave off the encroaching ice age detected by climatologists. Scientists are already worrying about what to do in a future era when the climate changes, and the earth overheats.

Historians recall the early part of the century with its exhaustion of some raw materials, particularly mercury, tin, lead, and zinc. But industry has compensated with new techniques of recycling and substitution. Scientists have invented "atom-transformers" – modern equivalents of the medieval "philosopher's stone", which they use not to turn base metals to gold but to turn the earth's crust into almost anything they want.

Work is done mostly by machines. For a time North American scientists advocated technologies that would enable 5% of the world's people to produce food enough for all. However, in order to avoid unemployment, the world government settled for an average of 20%. Industry is largely machine operated. For a time it seemed that there would be problems of too much leisure, but the world government met that by a moderate return from capital-intensive to labor-intensive production.

Public transportation is highly developed, but there are a considerable number of automobiles. The electric car never became popular, because of difficulties in developing compact and lightweight batteries. Cars are powered by hydrogen. Most people live in attractive environments and use sophisticated systems of communication, so there is less desire to travel than in past centuries. Homes are built according to styles originated long ago by Buckminster Fuller, and new building materials make housing plentiful for the reduced population of the world.

B. After the Nuclear Holocaust

Just how it started nobody knows. The evidence was destroyed in the war. The U.S.A. intelligence assumed that nuclear weapons landing on American cities originated in the Soviet Union, but that was only a guess. Rather than wait for more evidence, military authorities retaliated, and soon the world was involved.

Somehow due to a freak of atmospheric movements, most of Africa escaped the worst of the radioactive clouds that enveloped much of the world. America and Europe were desolate and remained so. Their cities are deserted. It is reported that a few people survived and still live there, but not much news reaches the rest of the world. China, it is said, is not so utterly destroyed as America and Europe, but it is in bad shape.

After the destruction Africa learned to live without any technological imports. The petroleum of Libya, Nigeria, and Angola was exhausted long ago. There are no airplanes, automobiles, or trucks. The great cities (Cairo, Nairobi, Dakar, Johannesburg) could not survive without petroleum, and they were evacuated long ago. Much of Africa learned without great trouble to adapt to rural tribal life.

The drought, which hastened the southward expansion of the Sahara desert, did great damage. Nobody knew whether it was due to natural causes or atmospheric changes brought about by the nuclear holocaust. For a time there was serious starvation, but Africa adjusted.

A few African historians of science are confident that they understand the technological skills of the late 20th century. They have annual meetings. They even talk about possibilities of inexhaustible supplies of energy through nuclear fission or fusion. But all their conversation is academic and theoretical. There is no adequate economic infra-structure to build the systems they talk about. Fires, for example, are mainly of wood. There is no steel industry.

Now and then somebody proposes the colonizing of Europe and North America. In these regions, there are said to be great deposits of coal and iron ores. It is conjectured that technologists could use raw materials from these under-developed countries as the basis for an industrial revolution. But it is rumoured that the radioactivity there is still too great for colonization.

The general assumption is that the future of humanity will be agrarian and that social units will be small and self-sufficient.

People occasionally speculate about the ruins of the abandoned African cities. There are legends and folk songs about the life that once prevailed in these cities, and a few traditional rituals are traced to their origins in urban life. Some people look on those cities with regret and envy. They appear to be a wonderful achievement, now denied to the present generation. Others look at them with revulsion. They represent a frantic way of life, combining privilege and oppression. It is rumoured that people in those cities scarcely knew their neighbours. Contemporary life looks better. Here people share their wealth and poverty. The arts belong to everyone. All join in common rituals and celebrations. Yet, say some, it would be pleasant if we could have just a little of the luxury that the bygone urban civilizations produced and used so badly.

C. The Nuclear Age

Looking backward, the choice now seems inevitable. Humanity, threatened with an energy shortage, had to choose the nuclear option. A few still disagree, saying that their ancestors should have opted for a simpler, less intricate technology and social order. But most people dismiss them as romantic neo-primitives.

Nuclear energy has been, in a way, a great success. But everybody also sees its problems.

Looking ahead, technological optimists and pessimists argue. The problems facing humanity are immense. But some say that we can expect continued research and new solutions. Others say that the "technological fix" has already shown its limitations and cannot assure rescue.

The world has become accustomed to a mix of productivity and austerity. Historians smile cynically when they report that people in the 20th century sometimes held up the U.S.A. and Western Europe as models for what they called the "developing countries". How, they wonder, could anybody, even in those intoxicated days, really have thought that India could ever have the number of cars per capita of the U.S.A.? Or how did aeronautical enthusiasts ever assume that half the human race would use air transportation?

Energy now comes primarily from "nuclear parks". The

world has 3,000 of these. The average one includes eight fast breeder reactors, producing 40 million kilowatts of electricity. To achieve this, humanity started in the late 20th century to build four new reactors per week. Now it is continuing that pace of expansion, but it is also building two more per day to replace worn out ones.

A large part of the world's economic activity is used to prepare and transport the nuclear materials, to dispose of the wastes, and to seal off the worn out reactors. Fifteen million kilograms of plutonium-239 are processed and transported each year. (Mesarovic and Pestel)

A worldwide surveillance system operates to prevent accidents, sabotage, and robbery. The system was accepted reluctantly, but after a few accidents and acts of terrorism the world decided it was necessary. There is constant monitoring of the nuclear parks, the waste disposal system, and the materials in transit. The guards are themselves monitored in a complicated system.

A central world authority enforces safety standards and supervises disposal of wastes. The salt mines of the U.S.A., the U.S.S.R., and China are valuable for waste disposal. Initially these countries refused to accept wastes from outside their borders, but eventually they agreed under pressure to take exported wastes.

There is an increasing feeling that the system may be getting unmanageable. Some see the answer in a successful fusion reactor. Every decade of the past century has heard promises that it will soon succeed, but its technology has proved elusive.

D.   The Planned Society

In the peaceable world of 2100 historians are still trying to understand how humanity made the transition from the insecurities of 1980. Some emphasize the counterculture within the industrialized societies of those days: the movement of people satiated with competitive industrialism to the point where they fled the cities and took up subsistence homesteading, either in family units or communes. Others point to the example of China, which established a self-sufficient economy by deliberately exalting human labour over advanced machine-labour and by requiring all citizens to spend some time in common labour. Still others accent the contribution of the Club of Home, an organization of technological and intellectual elites, which adopted the slogan, "Earth is our home," and predicted disaster unless humanity adopted life styles less destructive. A different group of historians say that humankind, haunted by fear of war and economic failure, was ready for any social system that promised security. Finally, there are those who say that a movement of prophetic mysticism, arising simultaneously in scattered parts of the world, let to the Refrev – that combination of reformation and revolution that modified human values and aspirations on such an amazing scale at the turn of the century.

Somehow the human race did what everybody had said would be impossible. It engaged in a large scale effort at social planning. There were, to be sure, disgruntled resisters. But a majority of humanity got caught up in the movement and persuaded or pressured the rest into going along.

What was most amazing was the combination of groups not used to working together – mystics and technologists, capitalists and socialists, to some extent even rich and poor. Some were lured by the vision of a new style of life, others moved by the threatened disaster accompanying old styles. The slowest to change were the political and economic leaders of the wealthy societies, who did not want to give up what they had. But they finally had little choice when a series of international cartels, formed on the pattern of OPEC (a 20th century Organization of Petroleum Exporting Countries) withheld exports, and when several threats from nuclear-armed terrorists in starving societies scared them out of their wits.

When E.P. Schumacher won the Nobel prize in economics, the fraternity of economists first gasped, then took a look at his work. Five years later a poll showed that 77% of professional economists took him seriously. Schumacher made peace with the

Club of Rome which he had been criticizing caustically. The United Nations sponsored an international conference that charted a new pathway for the world.

The globe was organized into a large number of economic units, relatively self-sufficient. International trade was drastically reduced, partly because it was becoming intolerably expensive in the face of petroleum shortages and price rises.

Use of nuclear energy was phased out, by international agreement, by the year 1995. All nations contributed through the UN to a research fund for development of alternative forms of energy, especially solar. Meanwhile there was a concerted effort to reduce energy consumption. Price mechanisms were already having that effect; a combination of rationing with tax policies tending to equalize wealth moved the process along. Travel and transportation – both international and domestic – have been greatly reduced by the year 2100. People cannot afford to travel much, and cities are designed to bring together opportunities for employment, living, and recreation. Sophisticated communication systems compensate in part for personal travel. Research funds go into public transportation systems. Private ownership of automobiles is forbidden, but people may rent cars (powered by methanol, made from sewage wastes) for special occasions.

There is some grumbling about a culture that constantly emphasizes economy, moderation in consumption, and the laborious recycling of resources. Some people say that creative enterprise is stifled, but others point to the resources that go into technological research, for example, in organic farming, in energy production, home designing, and utilization of substitutes for scarce resources.

Two years ago television networks in Europe and the U.S.A. began showing ancient films of the 1970's. Public authorities moved to suppress the films, saying that they were a pornographic titillation of illicit desires. There was widespread protest in the name of freedom of speech. Suddenly public authorities changed their policy. Instead of forbidding the films, they bought them and ran them repeatedly on the air. Their explanation was that the films, showing the intense frustrations in affluent societies, especially when alternated with films of wars in those days, were a sound education for people who compared the miserable past with the better present.

II. ETHICAL AND SOCIAL ISSUES

This exercise in imagination has its foolish aspects. Anybody looking at it ten years from now will ask why I picked these four sketches and omitted others. No doubt some are asking that now.

The point is to emphasize that social and ethical issues are concrete and historical. They depend in part on an empirical base. Before making a social and ethical judgment, I must ask: What are the real possibilities? What are the probable outcomes of different choices? What risks are involved?

But social and ethical judgments do not derive solely from "objective" data. They depend partly on visions, purposes, commitments. What is a good society? What is it to be human? Since to be human is to be a dreamer, what dreams are worthy to inspire our energies? Alvin Weinberg shows how scientific judgments keep edging into the trans-scientific. Wolf Häfele describes the situation of hypotheticality. Note that the realms of the trans-scientific and of hypotheticality themselves include two kinds of components. The one is a shortage of tested information that requires a venture into the unknown. The other is a conflict of visions as to what is good.

I shall here raise some issues and ask some questions. Some of my questions can, at least in principle, be answered by empirical data. Or perhaps I shall be told that the data are not available. I am accustomed to that situation. If I have my teeth X-rayed, I am balancing the benefits of the X-rays against the harm that they may do to my body. I do not know precisely the risk but I accept the best technical judgments I can get. If new evidence becomes available tomorrow, I may change my practices.

Other questions probably cannot be met by empirical data. After all the available evidence is in and after new evidence becomes available, I must still ask what I prefer for myself and my heirs. What kind of society will win my loyalty? For what goals will I make commitments, accept risks, and enter into sacrifices?

In this context of double uncertainty – factual and valuational – I shall examine a few issues.

A. Attitudes on the Relation of Humanity and Nature

The species called humankind is obviously part of nature, of an ecosystem. Yet humanity has qualities not shared by the rest of nature – ability to envision futures, make choices, bend environment to human purposes. Many attitudes are possible about the efforts of humans to take charge of nature. I mention two, chosen for the sake of contrast.

Hermann J. Müller, the famous geneticist, wrote in 1966:

Of course, we – that is humanity – will take our biological evolution into our hands, and try to steer its direction, provided that we, humanity, survive our present crises.

Have we not eventually utilized, for better or worse, all materials, processes, and powers that we could gain some mastery over?

Barry Commoner, on the other hand, coined the aphorism, "Nature knows best". Since that is obviously figurative speech, he explained his meaning (1971) that "any major man made change in a natural system is likely to be detrimental to that system."

Perhaps neither Muller nor Commoner is literally accurate. But they express two moods, two spirits, two wills towards human control over nature. Their statements are related to classical debates about humanity and nature: to the Chinese controversies between Taoists and Confucians, the Hellenic discussions of nature and custom, the Renaissance comparisons of nature and art. Is the highest human destiny to fit into a natural system, loving and enjoying it? Or is it the human privilege to replace nature with artifices? Or is the best plan to accept nature, cherish it, and yet transform it? There are analogies here to the classical theological discussions of the relation between nature and grace.

B. Risk

There is no life without risk. In biblical theology risk is inherent in the human situation, and security is an idol. There is nobility in some risks. Is it not better to take risks, even to sacrifice one's self for an ideal, than to cling to static security?

But some risks are rash and irresponsible. There is a special ethical problem in imposing risks upon others for the sake of ourselves. Perhaps there is a peculiar issue in imposing risks upon the unborn who are powerless to defend themselves. Informed consent has become an important ethical principle in medical experimentation. But future generations cannot possibly give informed consent to our acts that destroy possibilities for them.

Yet we know that we are beneficiaries of past risks. There was the risk of the first biological species that left the ocean and ventured to live on land. There were the risks of prophets who defied powerful authorities in obedience to a divine call. There were the risks of the heroes of liberty.

Likewise, we are victims of people who have taken risks for inadequate ideals, imposed their wills as conquerors on subject peoples, and shed blood to spread their creed and notion of good.

What are we to say of the ethics of the risks involved in nuclear energy? For an estimate of some of these risks most of us must simply look to scientists. They know more than the rest of us about the risks of radioactivity or of the malfunctions of systems.

When scientists differ, the rest of us must be perplexed. There is little truly "objective" data that we can comprehend. At this point I have an inclination to trust the disinterested scientists rather than those who have a stake in the success of their own ventures, simply because I trust myself most when I am disinterested. By disinterested I do not mean uncommitted.

When scientists move their evaluation of risks into the social and historical area, as indeed they should, more of us can get in on the discussion. We can learn something about what it means to build plants that must be kept safe, both when in use and when worn out. We can barely understand the meaning of nuclear wastes with a half-life of 24,000 years and a dangerous radioactivity for 200,000 years. We can begin to imagine the problems of guarding and protecting large quantities of dangerous materials under transportation across continents.

I am grateful for Alvin Weinberg, who in his advocacy of fission reactors, says: "But the price that we demand of society for this magical energy source is both a vigilance and a longevity of our social institutions that we are quite unaccustomed to." I appreciate his pointing out the importance of "surveillance in perpetuity". *(Anticipation,* No. 20, p. 13.)

My mind boggles when I try to multiply present risks by thousands, as I must do to grasp the problem, then to project it over the next century. If then somebody talks about thousands of years, I hardly know how to respond. My own country is now preparing to celebrate its bicentennial. I must ask, where has a stable society persisted for a thousand years? How often are natural systems stable for 10,000 years? In the face of such questions I find persuasive the warnings of Alvin Weinberg and Hannes Alfvén, different though their conclusions be. Without knowing for certain the answer, I must ask myself what present need justifies imposing a burden upon future generations for centuries to come?

C. Present and Future Values

Presumably the life of someone a few generations hence is as valuable as mine. But I am likely to favour mine, simply because it is mine. I am likely to favour my children's welfare over the welfare of my great grandchildren, because I know and love my children but have not seen my great grandchildren.

Furthermore, the future is unknown. In balancing present realities against future possibilities, I accent the present. The American baseball manager, Leo Durocher, always used his best available pitcher, saying, "Tomorrow it might rain." We know too little about the future to take it into account as forcefully as we do the present.

But at what point does unconcern for the future become morally irresponsible?

Most of us would take actions to prevent starvation now, even at considerable risk to future generations, who may learn how to handle the risks that we impose on them. But how much risk have we a right to impose on future generations for the sake, not of saving lives but of providing jobs for the unemployed – especially when unemployment is a function of an economic system that could be changed by other methods than stimulating production? How much risk have we a right to impose on the future for the sake of simple convenience and luxury now?

William Pollard pointed out that natural supplies of copper, tin, zinc, uranium, and even iron, which took "a billion or more years" in their formation, will be largely consumed in this century. He continued: "Men will look back on the 20th century as the age in which their spaceship was wrecklessly raped and her precious stores, accumulated over billions of years, were wastefully and thoughtlessly squandered." He added that fossil fuels, produced over a hundred million years, will be largely consumed in one century. He came to the conclusion that, if this human course should continue, "it can only prove to have been a cosmic blunder to have ever allowed man to evolve on this earth". (*This Little Planet,* pp. 60, 61, 64.)

Pollard requires me to ask whether the exorbitant demands for energy, required by our social systems, are part of a suicidal life style. To moderate that life style would be exceedingly difficult and, given present institutions, would inflict suffering on many defenseless people. Yet might it not be more ethically responsible to take the painful step now than to defer it until a time when catastrophe will be greater?

Of course, there is much about the future that neither I nor anybody else knows. It is silly to be overly confident of our projections, either of technological triumphs or of doom. But it is both silly and immoral to court disaster recklessly.

Herbert Butterfield, the distinguished historian and theologian, wrote:

> *The hardest strokes of heaven fall in history upon those who imagine that they can control things in a sovereign manner, as though they were kings of the earth, playing Providence not only for themselves but for the far future – reaching out into the future with the wrong kind of farsightedness, and gambling on a lot of risky calculations in which there must never be a single mistake.* (Christianity and History, *p. 104.)*

To agree with Butterfield is not to decide precisely what risks are admirable, what are irresponsible, and what are uncertain. It is to raise a question that most societies have not considered in any depth.

D. Economic Justice

Any form of power can be used for justice or for injustice. That is perhaps a cliché. Yet it requires constant re-examination. Was it C.S. Lewis who said that the supposed power of man over nature is in fact the power of some men, by managing nature, to control other men?

If we are here thinking about a general characteristic of human societies, it may have no special relevance to nuclear energy. But we should at least point out that the production of nuclear energy, by its very nature, requires a considerable concentration of wealth and power, possible only to a government or a major corporation. That is not in itself a reason to challenge nuclear energy, but it is a reason to watch closely its uses.

For example, we need to ask whether nuclear energy will be one of the many forces in modern society to increase the gap between rich and poor nations and between rich and poor in the same society.

We must also ask of nuclear energy, as of many other social phenomena, whether the costs are accurately assigned to those who get the benefits. Modern industry has often operated with deceptive accounting systems. It frequently turns a handsome profit by imposing hidden costs on innocent bystanders. For example, the costs of industrial pollution are paid, not primarily by the owners of industry or the consumers of its products, but by people who happen to live downwind or downstream from factories. The costs of the automobile are born partly by producers and consumers, but to a considerable extent by people whose neighbourhoods are torn apart by highways, even though these people cannot themselves afford to buy cars.

Any proper cost accounting for nuclear energy must reckon in not only the direct cost of production, but also the cost of experimentation, often born by government and not charged to the enterprise, and of surveillance now and for generations to come.

To raise this issue does not presuppose an answer that is negative towards nuclear energy. If it seems to do so, it is because that kind of question is so often neglected, either by technologists who plan operations or by industrialists who profit from them. Governments are theoretically in a good position to raise the question, but in practice they raise the questions that economically and politically powerful people want to raise.

E. Political Power

Inevitably we must ask about the role of corporations and of government in the production of nuclear energy. We can assume that in some societies corporations will do their best to persuade governments to pay a major share of the bills while the corporations reap the profits. So government participation, either in the actual production of energy or in the supervision of it, if government gives a voice to the people, may be a real advantage.

The cause for concern is the heightened power that may accrue to government in an age when people seem to find it increasingly difficult to make government responsible to their needs. Surveillance of nuclear operations is needed to protect people against accidental or malicious damage. To whom shall we entrust the surveillance? To a Richard Nixon and his White House gang, to a Vorster, to the military clique that rules Chile? The prospects are foreboding, even if we do not mention the generation of political leaders that perpetrated World War II. What happens to personal freedom in an age when technological operations are so intricate that concentrated authority must impose controls?

I hope I am not communicating the notion that I see a special demonry in nuclear power production or in technology in general. I think those who idolize or demonize technology are equally erroneous. I do say that those technological achievements that concentrate great power, when incorporated into social systems that are already inadequate, require us to ask social and ethical questions. Ideally, the intricacy and scale of nuclear fission might contribute to a heightened sense of the interdependence of humanity thereby raising human ethical sensibility. But it might equally well contribute to a heightened authoritarianism that bears down unduly on the weak.

The international political implications are equally important. If nuclear production of energy becomes commonplace, no nation can be indifferent to the safety standards of other nations, whether this concerns the hazards of production, of waste disposal, or of possible theft and destructive use.

The awareness of the international meanings of energy production might conceivably be an advantage in an age that has to learn how to outgrow certain kinds of nationalism. But the international corporations have shown us that internationalism does not necessarily bring power closer to the people. If a politically powerful nation and its weak neighbour each have a stake in the safety standards within the other country, one might hope for some kind of equality of provisions for inspection. But one might more probably expect an increased domination of the powerful over the weak.

Such political issues arise whenever a sophisticated technology enters into immature political systems, which seem to be the only systems we know.

F. Relation to Military Technology

Large scale production of energy by nuclear fission, I have already observed, means the multiplication of materials that can be used destructively by governments or by terrorists. In that way it obviously heightens the risk connected with widespread dispersion of exceedingly potent weapons. (See Frank Barnaby, *The Nuclear Age.*)

In another sense the relation of nuclear fission to military technology is important. The first major ventures in nuclear fission were part of a wartime effort to make weapons of unprecedented power. It is, of course, good that many of the people involved began immediately to look for peaceful uses of this scientific-technological achievement. It would be naive to reject a humanly helpful technology because it originated in a military effort. Humanity needs all the help it can get, and it had better not turn down any.

The question that must be asked is whether the relation of nuclear fission to weapons has vastly skewed governmental expenditures for research. The question almost answers itself. The immense research costs of developing nuclear fission would not have been authorized apart from military purposes. We must then ask what might have happened if the same money and intensity of effort had gone to devising other forms of energy or other social structures that might make life enjoyable without so much expenditure of energy. When we are told that technologies for using solar energy are not nearly as effective as nuclear technologies, is the reason inherent in the scientific nature of the case? Or does it result from the priorities that societies, for military reasons, have assigned to one kind of research?

There may be a curious circular argument at work here. That is, we may be choosing one form of energy production because the technology for it is more promising than for others; and it may be that the technology is more promising simply because we have chosen to emphasize that option.

G. The Meaning of Human Existence

On many social and ethical issues, I have said, my judgments will be influenced by the data that scientists produce. But finally we come to some judgments on which the poets and prophets, and indeed our own intuitions, have as much bearing as do scientific data.

Biblical faith testifies that humans do not live by bread alone. Yet the sacramental acts of Judaism and Christianity, the celebration of the Passover and the Lord's Supper, are acts of eating. This faith is neither materialism nor spiritual escapism.

Analogously we can say that humans do not live by physical energy alone, yet do live by the expenditure of energy. An energy policy, explicit or implicit, is part of every human life and society. The question is what human life style best expresses the profoundest meanings of life. Do the modern industrial economies, which some societies have achieved and to which many others aspire, best express the meaning of human life? Do they enslave or liberate the human spirit?

Curiously, Western civilization still lives to a significant degree from spiritual reservoirs of two ancient societies: the Hellenic and the Hebraic. Both ante-dated modern technology. Both were, by modern standards, societies of low consumption. They were certainly not ideal societies. They were always fighting their neighbours, and they were filled with political and economic injustices. Yet they showed some human possibilities that still illumine history.

If asked whether I want to return to their technology, I must say that I clearly do not. I am a person of my time and my culture. My world is a world of transoceanic air travel, central heating, telephone, television, hospitals, free institutionalized public education, availability of books. Such a world has formed me, and I would be uncomfortable in any other.

But whether this world can be sustained, I do not know. And how much of it is important to the pilgrimage of humanity, I do not know either. What am I, what is my culture, capable of giving up? What would we be better without? I don't know that.

What I think I know is that we had better entertain questions that we have never dared to face in this generation. To raise the questions, as I have done here, may seem to make me negative toward nuclear energy. That is partly because it is the business of ethics to raise questions that are often conveniently neglected, and the raising of questions may seem to be obstructionist, even if it is not.

But my aim is still to ask questions, to seek answers, and to be open to persuasion. So I would formulate my final question or pair of questions thus: Will nuclear energy offer the possibilities of a better human existence, bringing opportunities to many that have been available only to a few? Or would human societies be better with a simplified life style, involving reduced consumption of industrial products and bringing people closer to each other and to nature?

Our answer will in part be forced upon us as, for example, a reduced consumption of petroleum has lately been forced on many people. But our answer will be partly our choice. Our choices will be made in that bewildermen characteristic of humankind, where people choose with little knowledge of consequences and where events often seem to generate a momentum that governs choice. Certainly our choices cannot be infallible.

Yet it is important that we choose. To do less is to renounce our human gifts and potentialities.

REMARKS ON THE ETHICAL IMPLICATIONS OF NUCLEAR ENERGY

GÉRARD SIEGWALT

It is reported that at the end of the First World War the German Emperor Wilhelm II said: "Das hab' ich nicht gewollt" (I didn't want that). I think it is important for us to remember this affirmation when we consider what I would call the apocalyptic horizon facing us. We spoke during this Ecumenical Hearing of the risks, the dangers that surround us; the dangers from nuclear energy as it is becoming available to humanity which may not be sufficiently prepared for assuming responsibility for this new technology. In view of this horizon where catastrophes are impending dialogue is for us of greatest importance.

Since last year we have at Strasbourg University an inter-universitary seminar with representatives of the different disciplines which are concerned about the ecological problem. What has impressed us most in this seminar was the recognition of the fact that the different disciplines have each their own language so that the university resembles very much a tower of Babel. There are so many languages that there is great difficulty in finding a common language. It is the significance of a Hearing like this that we feel that we are more and more condemned to dialogue and pushed to find again a common language between the different disciplines. In this context I want to put forward the following four considerations:

The **first** consideration is an ecological one. Ecology has taught us that everything is related to everything, that there is an interrelationship between humanity, nature and matter. This means that a sectarian approach to reality – a unilateral approach to reality – will have effects on the other parts of reality that have been forgotten, and these effects may be very great. Ecology teaches us that there is also a limit to the earth, to the resources of the earth, to the possibilities of the earth and its inhabitants. We experience in a new way the finitude of man and earth. It has been urged that the regulative function of ecology be recognized by the different disciplines, so that the separate approaches to different aspects of reality be all referred to the one earth and to the one humanity on this earth. I think this ecological consideration has to be kept in mind when we approach the energy problem.

My **second** consideration is an epistemological one. Epistemology is the science of knowledge – to say it in a simple way: epistemology deals with the lenses through which we look at reality. If these lenses, these eyes through which we perceive reality are different, the reality as we see it will also be different. What is the epistemological assumption of the Western world? I have to be very brief: The Western world has followed the French philosopher Descartes who distinguished between the subjective reality of man, the *res cogitans,* the *ego cogito,* and on the other side, the objective nature which is object to human analysis and transformation. This is the Carthesian dualism which leads to what Marcuse calls in a special sense (but it is also true in a broader sense) the one-dimensionalism by which either the one side, the objective nature, or the other side, the subjective attitude of man, is absolutized. We live today in such a civilization where there is a gap, a split within and between human beings, a split between humanity and nature. The efficiency of this concept is very clear, we see it – it is our industrial and technological civilization, but we have also to see the weak basis on which this civilization is founded In his book, *Die Einheit der Natur,* Carl-Friedrich von Weizsäcker distinguishes between two illusionary systems *(Wahnsystem).* He describes one of the illusionary systems as being entirely illusionary, thus completely inefficient, and of little danger. But there is also another *Wahnsystem,* that is more dangerous because it is partially right. Weizsäcker asks whether our Western civilization is not such a *Wahnsystem,* that is dangerous because of its partial truth, and consequently its partial error. From this perspective, the nuclear problem we are dealing with is not only a technical one. There is also a philosophical and epistemological problem involved: What has already happened? What assumptions are made when we think in a merely scientific, objective way? I think this is the deeper issue we are confronted with: we should know that we are working on assumptions that may be questioned. This is not only a theoretical or even rhetorical deliberation. History, even recent history, teaches us that issues which are raised in a one-sided way provoke resistance from those aspects of reality that have been neglected or not sufficiently taken into account. But, though having been forgotten at one point, they will come back into the centre of man's memory at another moment in history.

The **third** consideration is an economic one. The concern for a sufficient energy supply which will satisfy humanity's energy needs is, in its place, a justified and good concern. The achievement of this goal involves planning. But there are two questions: (a) What are the motivations of economics? Are they also one-dimensional? Are they purely functional? Are they only related to economic man as a producer and consumer? Or is economics part of an integrated view of man and the whole world in which the economic concern is open to other concerns? (b) What are true and what are false economic needs? The distinction between real and wrong needs becomes necessary in a civilization where we have gadgets. Often we are overwhelmed by the delusions of basically superfluous needs. Hence the question: Is the assumption that electric energy will be doubled in ten years taken for granted? In the light of our quest for an examination of true and false needs, this assumption has to be challenged. The criterion for answering this question is the criterion of legitimacy. What is legitimate according to the view of humanity and according to the reality of earth?

My impression is that nuclear research is not condemned by these questions but that it should perhaps be used in a functional way, not so much as a producer of energy on a wide scale but for specific interventions where these are necessary. I refer to a parallel example: Using merely chemical fertilizers leads to a disequilibrium of the natural equilibrium. Chemical fertilizers may however be a complementary addition to organic fertilizers. As such they may be good, but used alone, they create a problem.

My **last** consideration is a theological one. I think that when we try to see together the different aspects we are faced with and to resist thinking in a sectarian, a unilateral one-sided way, then we are approaching in a new way the question of God. We are confronted with totality which transcends, which is greater than we are, but which we see appear. Thinking which is open for all the aspects that may be grasped by us, is necessary if we want to be in harmony with the whole of reality. Here we see a need for a theology of nature or of creation which enables us to relate every single aspect to the totality. When we try to do this – for it can only be an attempt – we may in a new way understand what the Bible calls the fear of God. In view of the totality of things we may understand the saying: The fear of God is the beginning of wisdom. This affirmation is directly in line with what I call the apocalyptic horizon in which we are placed. We seem to be experiencing anew today what the Greek called *moira* (fate). We realize that elements of reality which we have brought forward may follow their separate line, in their own way become autonomous, and finally demonic for humanity, leading to the disintegration of man and reality. There is then the invitation to what the Bible calls *metanoia,* the new thinking, and that means a thinking which is considering totality. Therefore the attempt to integrate thinking and dialogue as we try to have it here belongs to the worship of God as much as prayer does. Dialogue, integrative thinking, prayer, all are ways in which the fear of God is lived today, in which it can become real for us today. Through integrative thinking, dialogue and prayer, the *ecclesia,* the Church, is gathering, is occuring. A new community is being created and such a reality of *ecclesia* as we are experiencing it here is a hope for humanity and for the future.

SECTION 5

ECUMENICAL HEARING ON NUCLEAR ENERGY:

A Report to the Churches

SIGTUNA, SWEDEN – JUNE 24-29, 1975

The Central Committee of the World Council, meeting in West Berlin, August 11–18, 1974, requested the Sub-Unit on Church and Society to make an assessment of "the risks and potentialities of the expansion of nuclear power". After consultation with many persons active in the nuclear debate it was decided to convene a Hearing on Nuclear Energy bringing together nuclear scientists, scientists from related disciplines, technologists and politicians, as well as theologians and church leaders. The Hearing was held in Sigtuna, Sweden, June 24–29, 1975, at the invitation of the Swedish churches who helped to make the meeting possible. It seemed to be a worthy manner to help celebrate the 50th anniversary of the 1925 Stockholm Ecumenical Christian Conference on Life and Work.

The 30 participants in the Hearing Group reflected the wide range of views on energy problems in general and on nuclear energy in particular. Their widely diverging positions were outlined in a preliminary way in a preparatory prospectus for the Hearing which provided the themes for the meeting. In addition to background papers circulated in advance, a number of the participants were invited to submit papers on specific aspects of the issue which were discussed in plenary. The Hearing also benefited from the advice of a number of resource persons, representing various international agencies.

The World Council of Churches is especially grateful to the Dutch physicist Prof. H.B.G. Casimir who most competently chaired the Hearing, and to Dr. John Francis, a nuclear scientist from Great Britain, who served as Rapporteur and gave invaluable aid to the staff of Church and Society in the preparatory process.

The reports of the three Working Groups into which the Hearing was divided were discussed in plenary and, afterwards, revised and integrated into a single report by the officers and staff. This full draft report was circulated to all participants who have given their suggestions on the wording of different passages. It will be clear from the report that it is not a highly elaborated consensus document but should rather be seen as an attempt to describe areas of agreement and to interpret the divergent views.

The report of the Hearing is submitted to the World Council of Churches and its member churches in the hope that it will contribute to clarifying ecumenical thought and action on this decisive contemporary issue.

I. INTRODUCTION

HUMANITY'S ENERGY NEEDS

For thousands of years people had to rely on food and wood for energy; to add to the energy of their own muscles, they tamed and bred animals, and the stronger made slaves of the weaker. Wood was burned for warmth and for such chemical purposes as preparing food, manufacturing ceramics and metals. Later, muscular power was supplemented by the power of wind and water utilized by sailing ships, and by wind and water mills. By the end of the 17th century in the Western world, transport by water, the draining and irrigation of land, and several industrial operations had come to depend on such devices. Wood as fuel was gradually being replaced by coal.

Since then the world has seen an enormous rise in the use of fossil fuel and of non-muscular energy. One simple example suffices to illustrate this development: a hard-working man has to be well fed and in good condition to deliver one kilowatt hour a day, an amount of energy that can be bought today from an electric power plant for a small sum of money.

Many countries of Asia, Africa and Latin America are still characterized by an energy system drawing on conventional power sources. It cannot be disputed that these countries must satisfy the basic requirements of irrigation, food supply, and primary energy production; this will inevitably result in an expansion of useful energy supplies during the coming decades without indulging anything other than essential human needs.

On the basis of an analysis of energy consumption over the last hundred years it has been suggested that the world's energy consumption has been increasing at an average annual rate of 2% and that this growth rate will continue. It is clear that this rate of increase is far too low to redress the situation in the developing countries. Even a 6% increase would only lead to a fourfold increase in energy supply by the end of the century, a quite modest goal.

For the developed countries the situation is different. Several studies have predicted an annual growth rate of 6% for the rest of this century, but increasing attention is now being given to the possibilities of energy saving. A number of scenarios are possible. One, for example, is to cut energy growth by half which would still lead to a doubling of consumption in 25 years. Even this reduced rate of energy consumption imposes tremendous pressures on resources and environment. While many consider some expansion both unavoidable and desirable, a great deal can be accomplished by energy conservation and it is urgent that we investigate ways to save on use of energy.

However a growing number of people in the developed countries do not believe that this reduction is enough. They are convinced that a society living at a lower level of energy consumption than we have today would be better in many ways – more just, more aesthetic, more serene, more healthy and more stable. They advocate zero energy growth with a reduction in consumption as the ultimate aim. This would not imply a return to an archaic or pastoral economy for even if the energy consumption of developed countries were halved rather than doubled the per capita consumption would remain far above that in the developing countries today.

A reduction of energy consumption would involve change in the socio-economic pattern of industrial societies. It might be enforced by unexpected calamities, by authoritarian governments, or by a severe economic crisis. To achieve it in a more acceptable way would require a conscious effort strong enough to overcome the powerful forces of technological dynamism. Rising costs of energy and a growing awareness of environmental problems could help, yet it would be a slow process. It may be illusory to believe that a reduction in energy consumption in the developed countries will be reached in time to make possible the urgently needed increase in the developing countries. This emphasizes that the need to begin the transition is a matter of urgency.

If the lot of developing countries is to be improved they will need increasing energy supplies. Some of these supplies will come from fossil fuels which will be more readily available if the developed countries reduce their use of energy. And the developing and developed countries will need to have other resources of energy for the time when fossil fuels may become unavailable.

All countries are therefore looking for alternative sources of energy to fossil fuels. The immediately available option is nuclear power, but the development of nuclear power presents problems which are analyzed in this report.

II. PLANNING THE ENERGY SUPPLY SYSTEM

A. ENERGY NEEDS

Energy is essential both to sustain the high standard of living already attained in some parts of the world and to guarantee at least the minimum basic requirements of food, shelter, health and welfare to which the majority of the world's population justly aspire. Additional energy must be harnessed to meet the needs of the world's underprivileged majority, even though it must be borne in mind that although energy is essential for development, it is not the only requirement; socio-economic as well as institutional changes will be necessary if the less affluent are to reap true benefits from additional energy.

It is believed by groups in several developed countries that a more modest and waste-free use of energy is feasible without compromising the standard of living already attained – some even believe that the real well-being would be improved. However as explained in the introduction, it is unlikely that growth will be arrested at short notice, and it seems illusory to expect a reduction that could compensate for the growing needs of developing countries.

The total supply of primary energy which is expected to be available varies between countries depending on the nature and extent of resources and also on the level of socio-economic development. Few countries are totally self-sufficient with regard to all their present and foreseeable primary energy needs and most countries depend on a mix of energy sources. At present, fossil fuels provide for the vast majority of energy needs, particularly in the developed countries, but there is serious doubt whether these sources alone will adequately meet the additional requirements, discussed above, beyond the medium-term future (say the next 50 years).[1] Furthermore, because of certain economic and political factors, some countries are reconsidering their traditional energy supply patterns. Consequently, both the projected demand is being questioned and other sources of primary energy are being sought.

The forms in which energy is required for utilization depend to a considerable extent on the type of society concerned. The modern industrial society has developed in such a way as to be heavily dependent on electrical energy and, in the transportation sector, on liquid fuels.

B.   MEETING FUTURE ENERGY NEEDS

Until the last 15 years, the major sources of energy during this century were wood, coal, oil, natural gas and water power. Most countries have already developed most of their hydroelectric resources, although some, such as those of Southeast Asia, Africa and some parts of South America and Canada, still have substantial undeveloped resources. Before World War II, coal was the principal energy source, but since then oil and natural gas have largely replaced it. But both oil and natural gas, being depletable resources, are becoming scarce and the remaining reserves of both are becoming correspondingly more valuable and more costly. Geologists generally agree that the as yet undiscovered reserves of each probably at best do not exceed those already known or consumed. As a result, world consumption of both will probably peak by 1990 at 20% above the present level, and thereafter slowly decline to fall below half the present consumption by 2020. By then, the remaining reserves will have become so valuable that they will be too expensive to use for general energy production. Coal reserves will have to be used as a source of synthetic liquid and gaseous fuels for transport. Therefore less will be available to generate electricity. Waste heat from electrical power generation and solar energy will have to replace oil and gas for heating and cooling buildings. Rising costs will force increasingly stringent measures for conservation, elimination of energy waste and greater efficiency of its use.

Reduction of energy consumption in the developed countries would make the existing reserves last longer, but even if it could be realized it would not obviate the need for new energy sources.

Such energy sources will now be discussed under three headings.

1.   Nuclear Fission

Nuclear energy has been developed and used commercially

to meet an ever higher proportion of the electrical energy needs. (The present production of electricity by nuclear power is given in Appendix A.)

The advocates of this option are convinced that by the next century the required quantities of electrical energy at a price we can afford will have to come mainly from nuclear reactors. The availability of nuclear power based on current technology is dependent on world reserves of U^{235};[2] even optimistic estimates of the reserves of high quality uranium ore lead to the fear that at the present rate of deployment of nuclear plants, such reserves will be exhausted soon after the turn of the century. Unless other sources of electric power are developed we should then be forced to depend on mining of lower grade uranium ores possibly at great environmental cost and an energy input for the same electrical output much greater than the 5-7% presently required.

There are however alternative possibilities in the field of fission, viz. fast breeder reactors and thermal breeder reactors.

Fast Breeder Reactors. The liquid metal fast breeder reactor (LMFBR) is at the centre of an effort by the major industrial powers to increase the utilization of nuclear energy. The fuel for this type of reactor is usually a mixture of uranium and plutonium oxydes[3]; the heat is removed from the core by a circulating mass of liquid sodium metal and the primary circuit is not pressurized. In the breeding cycle the quantity of U^{238} that is converted into Pu^{239} is larger than the quantity of fuel (either U^{235} or Pu^{239}) that is burnt up by fission. If this development proves successful, practically all our uranium reserves can be used to generate energy – and these are 140 times as abundant as those of uranium 235, so that lower grade ores would not be needed so extensively. Prototype fast reactors have been operated successfully in the U.S.A., U.K., France and the U.S.S.R., although the breeding fuel cycle to recover the plutonium has not yet been fully developed or applied as extensively as would be necessary for the future programme. Fast breeder reactors offer the potential for electric power for the world's needs for centuries. The costs or risks of obtaining such energy from them are chiefly related to maintaining the safe disposal of the very long-lived high-level radioactive wastes which they produce (which are, however, the same as those from present reactors, the concentrations however being different). Furthermore, and equally important, society must guard against any diversion from the large quantities of plutonium in the fuel cycle to destructive purposes. The development and safe operation of such a nuclear power economy requires very considerable human technical and managerial skills. However the costs and risks, as described for fission are not significantly different for fast breeder reactors, but are most directly linked with the size of the nuclear power programme, particularly in relation to wastes and to plutonium transport and storage.

Thermal Breeder Reactors. These reactors burn uranium 233 and at the same time breed more of it from thorium 232 than they consume. The process has been proven experimentally in the U.S.A. and advanced engineering development may be as close or closer than that related to the fast breeder since the technology is similar to that of existing power reactors. Thorium is three times as abundant as uranium and a successful system of this type would greatly extend nuclear energy availability, although the rate at which new fuel is bred would be less than in the fast breeder. The waste disposal problem would certainly not be more difficult with thermal breeders, in some respects perhaps even less severe. Otherwise the costs and risks are comparable with those in the fast breeder economy.

2. Nuclear Fusion

For those who are not familiar with the advocacy of fusion energy, i.e. the combination at stellar temperatures of light nuclei to make heavier nuclei, there are several points worth making:
– The basic fuel, deuterium, is virtually inexhaustible and is ob-

tainable at relatively low cost. (However lithium, the second fuel, has to be mined and is not inexhaustible.)

- The by-product of the fusion reaction is helium, which is non-toxic, non-radioactive and hence causes no problem of radioactive waste management.
- There is no possibility of a runaway nuclear reaction.
- By comparison with fission reactors, there would be significantly lower quantities of radioactivity associated with the system.
- There is no possibility of diversion of strategic weapons material.

However, it will be at least five years and probably longer before the possibility of establishing a net energy producing fusion reaction has been demonstrated. Even if this is achieved, the engineering problems of developing a commercial power reactor using fusion still appear immense, and may well take another twenty years to overcome. It is, therefore, premature to base any concrete plans for meeting human energy needs on the use of fusion power until well into the next century (if ever). If, however, success were realized it would make a major contribution to meeting man's long-term energy needs. (It does not follow that the problem of managing the potential scale of energy production on a global basis is in any way reduced. The disruptive effects of major climatic shifts would undoubtedly place a serious restriction on the release of ever greater quantities of heat to the biosphere. The widespread availability and use of any new form of energy production will ultimately be limited by this consideration.) Although there is no major waste disposal problem properly speaking, large quantities of hazardous tritium would be generated to be consumed in the process; further the stored radioactivity of the reactor would be comparable to that of a fission reactor.

3. Non-Nuclear Possibilities

Solar-Electric Energy. Solar energy for heating and cooling buildings and heating water is practical and undoubtedly will come into increasing use at appropriate latitudes. It is also possible to generate electric power with solar energy by several methods, and a research and development programme to this end has now been launched in the United States. One way would be to use photovoltaic solar cells (semi-conducting devices that convert sunlight directly into electric power); this is done in the space programme, but the high cost of the devices used there is prohibitive. If a method could be found to produce a cell of this type cheaply in great quantity, a practical solar electric system could be developed. In places like the Sahara or Arizona deserts, where there is plentiful sunshine all year around, a 1000 MWe peak electric power plant would require 8 km^2 of such solar cells with 12% conversion efficiency and its daily average annual power would be 250 MWe. To provide larger electrical capacity would require more land area. This land area requirement coupled with the problem of energy storage will restrict the worldwide usefulness of solar energy for electric power generation, regardless of the success of the research and development now underway.

There are other ways of harnessing solar energy. Differences between surface and sub-surface temperatures in the oceans may represent a promising energy source. Wind energy is also a form of solar energy. Further it should not be forgotten that all agriculture and forestry transforms solar energy into chemical energy; better crops might be developed (See Appendix B for non-nuclear energy programmes in developing countries.).

Geothermal Energy. The deep rocks of the earth's crust are hot and constitute an immense energy universally available. A technique for tapping this energy resource for electric power generation has been suggested. A hole would be drilled to some 5–10 km and a large cavity of broken rocks formed by hydrofracture or a nuclear explosion at the bottom. Water would then be admitted to the cavity and hot high pressure steam withdrawn to a steam turbine generator on the surface through an auxiliary drill hole. To many it seems likely that the cost of creating such a system divided by the amount of electricity generated from the steam produced will result in a very high cost per kilowatt hour; but this is disputed by the advocates.

C. COMPARING THE DIFFERENT POSSIBILITIES

Unfortunately it is not possible at the present state of knowledge to make a strict comparison between the availability of nuclear energy systems and the potential development of non-nuclear sources of energy production. Such comparison must await the results of the major research and development programmes that have been initiated in different countries. But on the basis of past experience, and the very long time lags encountered in the introduction of major energy sources, the implementation of any technological breakthroughs cannot be anticipated before 1985 at the very earliest even on the most optimistic assumptions.

Nuclear fission is at this moment the only technically and industrially developed possibility to add to or to replace fossil fuel for large-scale production of electricity at a reasonable cost price. The decision to introduce ever larger nuclear plants all over the world is a fateful one. But the advocates of nuclear energy point out that the alternative might be as fateful and lead to severe social and economic disruptions and upheaval.

Opponents of nuclear energy are optimistic concerning the non-nuclear energy sources, and many of them would even be willing to face such upheavals. They point out that the industrial-technical civilization we have built, with its insistent demands for exponential growth, is a trap into which we have all been quite unwittingly led. Because of the finitude of the earth's resources, this growth must some day end. The whole of humanity must somehow find the wisdom to imagine a different life-style and mode of existence on our planet earth, and set about working toward it in a way which will encourage a significant change in the dynamics of our economies as these are presently constituted.

As a matter of fact, there is within several countries growing public questioning of the decision of authorities in favour of the nuclear option. This has already led in some to a slowing down in the rate of increase in installation of nuclear capacity and a decision to reappraise the advantages and risks in the light of further experience, e.g. in the Netherlands, Sweden and the United Kingdom. Furthermore, certain technical, political and social factors, to be discussed below, are leading concerned groups in some countries to question the wisdom of using nuclear energy at all. They believe that society could be organized in such a way as to guarantee an acceptable quality of life for all by an effective use of energy resources other than nuclear energy. It is argued that the most appropriate technology should be employed in the most efficient way to meet the specific needs of people under their particular circumstances: local resources should be used wherever possible, thus reducing dependence. It is further held that people should strive to organize their society and the technology they use in the way which is most harmonious with nature and thus maintain a proper ecological balance.

D. ISSUES FOR FURTHER CONSIDERATION

The Hearing Group was not of one mind concerning the desirability of the use or non-use of nuclear energy, but wishes to call attention to a number of important issues. The consequences of either including or rejecting the nuclear option in providing for the energy needs of a particular society should be considered from all angles. The Hearing Group has concentrated essentially on the issues related to the use of nuclear energy, but those wishing to pursue the non-nuclear option should likewise take full account of its consequences.

The following issues are raised in relation to the use of nuclear energy:

1. Most of humanity can foresee the dangers of misusing nuclear technology. This relates in particular to the nuclear weapons field. Many scientists and intellectuals, aware of these dangers, have not been able to prevent such misuse in countries which have developed weapons; when they are themselves actively en-

gaged in research and development in the nuclear energy field they may be faced with hard ethical dilemmas. But even if many scientists were to disengage from active research in the field of commercial nuclear power development, this would be unlikely to halt the widespread implementation of planned expansion in many countries.

2. The introduction of nuclear energy on a large scale will help to reduce the gap between industrialized and developing countries only if there are also changes in the social and political structures. Otherwise the use of nuclear energy could even widen the already existing gap and magnify the power imbalance between developed and developing countries, as well as between rich and poor within a country. The use of nuclear energy should be geared to the search for a new international economic order which will further international justice and peace.

3. A group of countries and transnational companies have been pursuing a policy of increasing the price of liquid fuels. This is of medium-term benefit to the oil producing countries and the companies, and has created favourable conditions for them to embark on research and development of alternative sources of energy, particularly nuclear energy. Thus, the world is in a transition period toward the post-petroleum era and this has introduced a number of new elements into the international situation. Oil producing countries are benefiting, gathering considerable revenues; countries already disposing of advanced technology and capital are investing in the development of new sources of energy. But for a considerable number of developing countries which have neither oil reserves, nor advanced technology, nor capital, this increase in the price of fuel presents only disadvantages. They are obliged to allocate huge amounts of money for their energy supply programme, which they could otherwise well use in other sectors of their development. This further widens the gap between the rich and the majority of the developing countries and tends to preclude the latter from establishing and implementing energy options best suited to their own conditions. They become increas-

ingly influenced by the energy options of developed countries, even more dependent on international or bilateral aid, and subject to the influence of the multinational companies.

The Hearing Group was aware of several other important problems, but it was not able to discuss them fully. They are mentioned here as a stimulus to others who will undoubtedly advance the debate on nuclear energy on some future occasion.

1. Modern large-scale technology involves the risk of causing a rift between technology and the wholeness of humanity and nature. Therefore the social and physical sciences, as well as the life sciences, should make a common effort to study the consequences of nuclear energy.

2. Decisions concerning the development of nuclear technology should be made primarily at a national level without the interference of other governments and especially without the interference of transnational companies or any other large economic unit. The fact that technical information, materials and apparatus of an advanced nature can often be obtained only from such firms or units may easily lead to an undesirable form of indirect interference. In this context, international agreements, in particular on nuclear safeguards and operational safety measures, should not be considered as interference. The safe use of nuclear energy calls for effective and enforceable international agreements.

3. Notwithstanding the national autonomy emphasized in (2), it is essential that nuclear technology be shared. The idea that one group of men or nations is capable of deciding which people or nations are entitled to nuclear knowledge is unacceptable. At the same time, public consultation and debate should take place in and between countries prior to any country embarking on or expanding its nuclear power programme; it should not be the exclusive right of governments or of the scientific community to decide in these matters. We must also take into consideration that no nation has by itself developed nuclear energy. Even more important

is that the development of nuclear energy by one nation may have detrimental consequences for others. This has been clearly shown by experience.

Finally, the Hearing Group calls attention to the need for further examination of the following consequences of either accepting or rejecting the nuclear option:

1. The socio-economic changes required by nuclear and other energy systems. For example, the use of nuclear energy in an acceptable manner for the benefit of society as a whole will require fundamental changes in the present socio-economic system and certain safeguards against hazards and sabotage, etc. If nuclear energy is not accepted, institutional arrangements will have to be made to ensure maximum benefit from other energy sources. In all circumstances there will be need for new emphasis on the elimination of unnecessary waste of energy.

2. Planning of further research and investment to develop nuclear and alternative energy sources on a wide scale. Although considerable research and development has already been done in the field of nuclear energy, its exponents consider that much more will be required to attain a level of supply which is virtually inexhaustible. For a non-nuclear society, considerable further research and development will be required to exploit alternative resources. (See Appendix C for a description of future prospects for research and development of energy resources.)

3. Careful investigation is needed into the effects on people and their environment of nuclear energy production using the full range of scientific and engineering skills. Most people are aware of the essential need to contain harmful radiation and toxic substances associated with the use of nuclear energy. However it is not only systems of producing energy which require great care, but to a greater or lesser extent anything that man does – agriculture, travel, industry, the burning of coal and oil, etc. disperse harmful and toxic substances. Every system of energy production inevitably has an impact on the environment, and before any new

sources are exploited, full account should be taken of their long-term environmental consequences.

III. RISK ASSESSMENT AND THE CRITERIA FOR PUBLIC ACCEPTABILITY

A. Risk Acceptance in a Technological World

The safety of a nuclear energy supply system will depend on the excellence of its design, operation and maintenance, and on an overriding desire to achieve the safety of people working in it and of the public throughout the life of the system. Under these conditions, the chance of a serious accident is likely to be small, perhaps smaller than the risk to the public from some other industrial activities. Such risk as remains will be reduced progressively through experience in design and operation supported by relevant research and development programmes.

During the period of industrial growth many accidents have occurred – in transport, mining, factories, etc. The lessons learned through these accidents led to the slow elimination of many of their causes until a low and tolerable frequency was reached.

In the modern development of very large, complex and potentially dangerous plants, such as those producing petro-chemicals, pharmaceutical products, large aircraft, and nuclear energy, we cannot wait for accidents to happen in order to improve designs; consequently techniques for the recognition of potential accident situations have been improved on the basis of relevant experience of failures in machines and in man-machine interactions. There is good evidence that through these methods accident possibilities may be circumvented by design or by care in inspection and operation.

There is no infallible way of foreseeing all possible types of failure. Present techniques are good, but they are not yet widely applied and the engineering training in most countries is notably weak in providing the necessary technical insight and discipline to reduce the risk of accidents. At the same time there has been no international effort directed towards the full scale simulation of accident conditions on a major nuclear power plant. While engineering confidence in the design and assembly of reactor safety systems may be high, these systems remain largely unproven under the predicted failure modes for different types of reactors.

The "maximum credible accident" study has been useful in the first stages of nuclear plant development. There has been a fundamental difficulty in defining the Maximum Credible Accident – it cannot be based on a logical premise as there is no framework of experience to establish what is the maximum that is credible. The current trend in most nuclear plant development is toward a measured assessment of all failure modes, or those accident chains which are considered important – the extent of the analysis depending on the consequences foreseen.

During such an assessment – which will continue during the operation of the plant – an agreement is reached between the responsible bodies as to the design basis to be used to reduce the likelihood of the accidents occurring.

There is a growing awareness of the importance of common-mode failure[4]. Good design and independent review of design and installations will gradually reduce but not eliminate the chance of unforeseen failures of this type.

The importance of safety and reliability of shut-down systems (i.e. the systems designed to close down the reactors under emergency or fault conditions) – or other systems such as emergency cooling and emergency power – is widely recognized. There have been notable improvements in design in all reactor systems and a move toward increased redundancy and diversity, towards the use wherever possible of different mechanical systems and different measuring devices using different components to reduce the chance of error. All systems are subject to tests of part or all of the systems at predetermined intervals under conditions under which they may have to operate.

Certain criteria have been established to provide a rough and ready guide to utilities in their early search for nuclear sites. Many factors are involved – foundations, water, power distribution, the proximity to people and to other industries, seismic faults, etc.

The siting criteria for nuclear power plants have been progressively relaxed over a period of years to achieve the full economies associated with the minimizing of transmission and distribution costs for the electricity produced. New reactor sites are now allocated in areas of low to medium population density although there has been a slow approach to major cities. This has varied between countries as there are significant differences in mean density and in local variation in density. Within each country siting could not always be based on an actuarial risk, but was determined rather by balancing the desire to be remote from major cities against the difficulties of using remote sites for industrial purposes, the need to shorten supply lines, to provide labour, etc.

Much progress has been made in assessing the risk to people or the environment from nuclear reactors, and the results of these assessments can and should be made available to governments and people who may be affected. Whether the predicted low risk of harm is realistic or not depends on the care and continued discipline applied to ensure safe operation.

Chemical reprocessing of spent nuclear fuel is a difficult operation, but many of its short- and long-term effects – such as those of discharging tritium or krypton into the atmosphere – can be reasonably assessed. It is the declared intent of reprocessors to control these by absorption and storage. Small spillages will cause moderate to severe local problems as in other chemical processes, and great care is required to ensure that these effects are not spread more widely as through rivers, underground water, etc. The main problem however is the high radioactivity content of such plants and the medium and high-level wastes produced.

The long-term disposal of wastes – particularly those which are highly active and have a long lifetime – remains a problem. How it can be ensured that they will not be a perpetual source of harm for 1000 years or more to come is yet to be discovered. The process for rendering wastes insoluble through glassification[5] is still in the early stages of development, although there is good reason to believe that this can be carried out on the appropriate scale. However major application on a large scale has not yet been attempted! Satisfactory results have been obtained from experimental disposals monitored since 1959.

This problem of waste disposal is a particularly critical area for international collaboration and agreement, even though the temporary disposition of waste is at present being dealt with inside national frontiers. Problems of earth movement, dissolution and migration of nuclear wastes are essentially global in character.

It is not yet possible to say whether good solutions will be permanently available; neither can one say that this will not be the case.

Standards for transporting irradiated fuel, plutonium and other radioactive isotopes, for containers and testing procedures have been evolved by the International Atomic Energy Agency with other international and national bodies over the past 20 years. For the highest risk materials, containers are subject to severe tests equivalent to high speed collision followed by intense fire. Although accidents may and no doubt will occur, their effects in a well ordered society are most unlikely to hurt people and will cause only local disturbance. There is greater concern about the possibility of wilful diversion and misuse by criminal or terrorist organizations.

B. Public Acceptability

Public acceptability should be an important criterion relating to the utilization of nuclear energy in determining public policy. However, in most democratic societies an open discussion of the issues involved is only beginning, and public acceptability is therefore hard to assess for the following reasons:

Any meaningful decision depends on a weighing of available alternatives. For example, a society might in a general way favour a reduction in pollution, yet simultaneously refuse rationing or taxation designed to restrain consumption. Public acceptability is therefore almost meaningless except in relation to specific decisions.

There are many publics within a society and there will no doubt be serious conflicts of interests. There may be a widespread general feeling that the society would be better off with reduced usage of private automobiles. Yet corporations and trade unions manufacturing steel, rubber, glass, and finished automobiles might favour increased production of cars. Likewise there are many physical planning and locational factors that provoke public reactions, e.g. many people want increased energy, but do not want a hazardous or dirty power plant in their neighbourhood. In this connection the right of minority groups is posed in a new way.

Through oversight the public might assume some risks that it would never deliberately choose to accept. For example, most societies would be against the number of casualities associated with the motor car. The same society might, however, by a series of decisions about speed, unintentionally accept a high casualty rate.

All social decisions have an element of ambiguity and this element is increased in a complex technological world. When competent technical opinion diverges, public acceptability often becomes exceedingly difficult.

These difficulties point to the importance of and need for socially responsible ways of making public decisions. They require an informed public. There is need also for governmental measures to distribute the costs and gains of innovations (or the rejection of innovations) more justly than is generally the case. But the difficulties of the problem cannot be an excuse for evading it.

Here we meet an intrinsic difficulty. The plurality of publics and the conflict of interests represent also a plurality of value systems and conflicting understandings of justice based on incompatible visions of the good. The churches could contribute substantially to discover the truth in the statements of the opposing interests in such conflict situations; they should also support the rights of people to essential information.

We feel it is important to stress the value of a procedure such as the one adopted in Sweden (Appendix D), where an effort was made by government to inform the public on the risks and benefits of nuclear energy. In this way it becomes possible to hope for a high degree of public participation in the discussion of even a technically complicated issue which may have very important consequences for society.

C. Problems Arising from the Diversion of Nuclear Materials

The Hearing Group is concerned over the possibility of the clandestine diversion of plutonium for the construction of a nuclear explosive. The danger is not confined to governments – nongovernmental groups or even individuals may construct and explode, or threaten to explode, nuclear devices. Plutonium could be stolen; it could be ransomed; for various motivations a group may seek to obtain plutonium in order to use it as a nuclear threat.

This year, the world's nuclear power reactors will produce about 25,000 kg of plutonium; in 1980, they will probably produce about 80,000 kg. The exponential increase in the amount of plutonium produced may continue for the rest of the century.

International safeguards, as at present applied, can only **detect** the diversion of nuclear material, they cannot physically **prevent** it. Moreover, these safeguards are primarily designed to detect diversion for nuclear weapon production, with the hope that early detection will deter such diversion.

While the International Atomic Energy Agency can handle its present obligations satisfactorily it is less certain that it will be able to cope with the greatly increased use of nuclear energy predicted for the next decade.

Under these circumstances the need for adequate physical protection of plutonium becomes critical. This task must, at present, be performed by the national control systems since most states today would resist this being taken over by an international agency on the grounds that it would conflict with national sovereignty.

The greatest need for protection is from the time the plutonium leaves the reprocessing plant until it is eventually burnt as reactor fuel. The protection must prevent both theft by persons employed in the nuclear facilities and theft by outsiders. Perhaps the most critical time would be when the plutonium is being transported between facilities; and, in particular, when it is being transferred from one method of transport to another. Once again, the problem is greatly complicated by the considerable distances that may exist between the establishments housing the plutonium. A great many problems will arise if an absolutely foolproof system to prevent hijacking is required.

By the maximum use of the appropriate available technologies plutonium could probably be made relatively secure, at least in the short term. But two problems immediately arise. As the amount of plutonium to be protected becomes very large the cost of adequate protection may become prohibitive. The measures which will be necessary may become cumbersome and too complicated for easy implementation. They may also be unacceptable in a democratic state.

Such fears have been expressed in relation to an expanded nuclear programme by Hannes Alfvén who holds that: "The high complexity of the nuclear systems in combination with the risk for catastrophes give rise to unprecedented requirements for a standard of excellence in design and operation and in reliability of components and personnel. It seems doubtful whether these requirements can really be satisfied when a large number of reactors come into operation in many countries having widely varying experience in high technology."

For these reasons (and because many countries may not adopt fully adequate national control systems), an alternative solution to the problem is suggested by some. One possibility would be to forego the future use of nuclear power altogether. Another solution would be to put all processing plants, plutonium stores, and reactor-fuel fabrication plants under regional or world-wide international supervision.

There may be other possibilities, but summarizing the alternatives we find that one group believes that the process and controls required to safeguard nuclear installations will become so complex and difficult to administer and the chance of failure by oversight or mischief become so great that nuclear power should not continue in the present state of the world.

There is however the other group which believes that while the difficulties are great, the need for energy is also so great that a degree of national and international discipline will evolve to ensure that risks are kept within bounds and to ensure the rapid promotion of all the techniques required to understand and adequately control the problems now foreseen.

D. Concluding Observations on Public Responsibilities in Risk Assessment

We note that international boundaries (and to a much lesser extent administrative boundaries within a country) can lead to inadequate weighing of interests in the siting of nuclear installations of all kinds. Those on the "wrong" side of the boundary are nearly always inadequately informed, inadequately heard, and have no influence or only inadequate influence on the decisions taken.

In the case of nuclear accidents, such boundaries may hinder rapid and effective protective action for the persons that could be affected, actions which may be required to control the accident or its effects (e.g. collective action between countries may be necessary in the event of a major accident occurring in close proximity to a frontier or boundary region).

We urge all authorities to fulfil their moral obligations to all concerned by ensuring that the boundaries of their jurisdiction do not lead to less responsible decisions and effective actions than if these boundaries had not been present.

The benefits of nuclear power may well accrue to the local or national community, while many of the longer term risks (and disadvantages), i.e. through radioactive wastes, plutonium diversion, etc. can affect a far wider area. This introduces a bias which may lead a community to be more favourable to nuclear development than if their area of responsibility covered all people involved.

Some risks and responsibilities will fall on generations yet to come. This leads to the difficult problem of how to balance present benefits against future risks and burdens.

IV. RADIOLOGICAL HAZARDS AND OPERATIONAL EXPERIENCE WITH NUCLEAR POWER PROGRAMMES

A. Normal Operations – General

In our context, large-scale nuclear power programmes are taken to include all elements from ore mining and milling through fuel fabrication, reactor operation, waste removal reprocessing and storage. There are special problems related to the mining and milling area where historically different standards have pertained at various times.

We are concerned with both workers in the industry and the populace at large. The current maximum permissible radiation doses (MPD) as defined by the International Committee on Radiation Protection (ICRP) are 5 rems per year for occupational workers and 0.5 rems per year for individual members of the populace[6].

Experience with the various steps in such programmes, including the nuclear fuel industry, nuclear power reactors and waste handling, has shown that there is no difficulty in keeping the radiation doses for occupational workers within the ICRP recommended limit. In fact, the doses have generally been much less, and, if necessary, it will not be difficult to accept lower limits for both occupational workers and the public, even with a widespread availability of nuclear power. However, it is to be noted that the average radiation dose received by workers in power producing units and reprocessing plants has been rising rapidly in some cases during the past years. Some countries have adopted much lower levels as national standards, not because radiation risks from low doses have been found to be greater since the ICRP recommendations were published in 1965, but because of the recommendation of "as low doses as practicable". Thus in West Germany a collective dose rate of < 1 millirem per man.year to the public is recommended as a guideline for the releases from the nuclear power industry. In Sweden the recommended limit is a collective dose of < 1 man.rem per megawatt year nuclear electric power; this corresponds to 10 millirem per man.year for 100 GW electricity installed and 10 million people. It should be observed that these values are about 10 to 100 times lower than the average radiation dose received by the public either from medical X-ray examinations or from the natural radiation background.

At the same time, the surge of growth of the nuclear industry is the source of concern with respect to the radiation dose level for occupational workers since the number of exposed workers will rise sharply (possibly reaching several hundreds of thousands), and since the search for minimum cost of operation will tend to lead employers to consider that the MPD can be delivered to every worker. For these reasons, as well as the fact that more and more temporary workers are used, we express concern about the control of radiation dose. It appears highly desirable that an international or at least supra-national, regional nuclear passport be created for every worker, recording his/her radiation dose history. A similar system already exists, for example, in Switzerland.

Many steps of the nuclear fuel cycle involve releases of small amounts of gaseous or liquid radioactivity into the environment, as well as some direct radiation to the occupational workers. (The total doses received as a consequence of this have been discussed above.) Considering all sources of radiation from a nuclear reactor system, including reprocessing, it is worth noting that releasing ^3H

and ^{85}Kr to the atmosphere may become the main source of radiation dose (from the nuclear power industry) to the population. Techniques exist at present for collecting ^{85}Kr. Such techniques must be applied in future reprocessing plants. The technique for collecting tritium is more complicated, and it must be expected that fairly large amounts will be released into the environment, most probably in the form of water. The difficulty of retaining it may become a limiting factor for the nuclear industry.

Because of the rising costs of safety procedures in uranium mining and fuel reprocessing plants it is to be expected – and this is already happening in the case of mining – that these industries will tend to be preferentially located in those countries with less effective controls. An international control of the safety procedures throughout the fuel cycle would therefore be desirable.

We are not aware of any significant determinable effects of radiation on either workers or populace through normal operations of power reactors (including subsequent processing and waste disposal). At the same time, the rapid growth in nuclear power programmes will almost certainly mean that an increasing number of people will be exposed to an increasing fraction of their MPD. Pressures will therefore increasingly exist to reduce radioactive releases to remain within permissible dose levels. However, there is still great uncertainty in the experimental findings relating to the risk of either somatic or genetic effects of low dose rates of multiple small fraction exposures. The policy so far has been to err on the right side. Still this problem calls for greater efforts in both the experimental area and in the gathering of relevant biostatistical data. It is prudent to assume from current knowledge that no threshold of radiation dose has been established below which there is no effect. At the same time no unsafe effects have been noticed when working with the currently established ICRP dose limits for the last three decades and adapted to local conditions. There must continue to be evaluation of any possible relationship between the lifetime or cumulative dose of groups of workers (and/or of the populace at large) and specific physiological effects. This must pertain both to radiation arising from nuclear operations, from natural sources and from medical applications. The prospect of the development of medical radio-protective measures can aid, for example, in meeting needs in specific urgent accident·or rescue situations.

B. Fast Breeder Reactors

In the case of the liquid metal fast breeder reactor (LMFBR) system, there are a series of major unresolved issues relating to instrumentation and safety. While these are highly technical, they may be worth quoting:

1. The adequacy of instrumentation to detect fuel pin failures, coolant channel blockages, and local fuel melting, without any serious time delays;
2. the nature of molten fuel – coolant interactions and the possibility of local accidents propagating through the reactor core;
3. the adequacy of mathematical models and computer codes to place realistic bounds on the explosive potential of a fast reactor nuclear excursion, and the consequences which might then follow: i.e. the release of plutonium and all the high boiling point fission products.

It has been predicted by some experts that the risks from a severe nuclear accident on a light water reactor, may be of the order of 10^{-6} to 10^{-7} per reactor per year, and this level of probability is comparable with the risk from routine releases in the same plant. If this low frequency was maintained, the dispersion of radioactive material following accidents will then be less than routine releases calculated on a cumulative basis (e.g. severe accident releases may be 10^7 curies I-131, 10^6 curies Cs-137). It is the release of this quantity of radioactivity in one place at one time that becomes the critical factor in the risk calculations. This relationship will hold true for other types of nuclear plants including the liquid metal fast breeder reactor, although it is not possible to show at this time that the accident probability will be acceptably low, i.e. below about 10^{-6} per reactor per year. (10^{-6} per year means one chance in 1,000,000 per year of operation on each reactor.)

Under normal operating conditions, the special hazards of a fast breeder reactor are mainly related to the much higher fuel coolant (in the case of sodium) and fuel with high plutonium content.

Events such as:

- Sodium-water reactions and the associated fire hazard;
- the introduction of moderating materials into the core with corresponding changes in reactivity;
- the blockage of narrow coolant passages, particularly in the vicinity of heat transfer surfaces within the fuel element bundles;

are different in kind from those in thermal reactors and necessitate very specific design considerations. In addition, under accident conditions there is the possibility of a re-arrangement of the core into a more reactive (eventually prompt critical) state. A subsequent vapour explosion which terminated the reaction was one of the special phenomena first associated with the sodium-cooled breeder reactor. More recent studies have shown that other non-explosive ways of terminating prompt criticality exist and they are being actively investigated at the present time. The situation can be summarized as follows:

The existence of a significant Doppler effect in large fast reactors is an extremely important feature[7]. In large fast reactors using ceramic fuel containing a high proportion of U^{238} it was found that the fuel lifetime could be extended and at the same time a mechanism was found to exist which could instantaneously reduce the effect of inadvertently adding large amounts of reactivity which could cause the reactor to become "prompt critical" or uncontrolled. In a typical large sodium-cooled fast breeder reactor, for example, it is possible to imagine a very improbable sequence of events in which some accidental re-arrangement of core geometry causes a sudden increase in reactivity while the reactor is running at full power. If we further imagine that none of the many automatic safety devices function so as to insert the safety rods, then the fuel temperature will begin to rise rapidly until the compensating effect of the Doppler process counteracts the added reactivity when the temperature rise is checked for a brief period at some level below the point when fuel damage would occur. In this way, the reactor's response to the initial disturbance is controlled and the difficulties of limiting the effects of the energy release are greatly mitigated. The major factor in this hypothetical accident which prevents the reactor from becoming uncontrolled is therefore the Doppler effect. Confirmation of the studies in progress should give added assurance that plutonium will not be dispersed outside the immediate environment of the reactor under fault conditions.

C. Waste Management

Radioactive wastes are produced by all operating plants in the nuclear power industry. The basic system adopted in waste management is the dilution and dispersal of the low activity wastes, and the concentration and containment of high activity wastes. The latter, which emanate from the fuel-reprocessing facilities, constitute about 99.9% of the total radioactive substances produced. They consist mainly of fission products, with some transuranics (actinides). The radioactivity content may be up to 10,000 curies[8] per litre. They are stored first in underground stainless-steel-lined concrete tanks with double containment, and cooling for the decay heat. However, tank storage for such wastes is an interim measure, and all countries faced with this problem have long-range programmes. These are: (1) Storage for a finite period, e.g. 5 to 10 years, in tanks to allow for decay cooling. (2) Solidification in a non-leachable stable material such as glass. (3) Storage of fixed wastes in an interim vault under controlled conditions, with cooling to dissipate heat and allowance for decay time. (4) Permanent disposal (if it is necessary to remove the wastes from storage) in a geological environment stable over geological times and with no water ingress possibility. Formations such as deep abandoned salt mines and basaltic and granitic formations have been investigated by different countries.

The problem of transuranic elements in high activity wastes has been of concern to many because of the long-term control requirement. One solution advocated by some is their separation from the bulk of the high activity waste, separate treatment in special reactors, and conversion to less hazardous isotopes. There is need for practical demonstration of such methods, which would likely increase the cost of waste management. Experience so far indicates that high level activity wastes can be managed safely without separation of the transuranic elements, though review will be needed if and when fast breeder reactors come into operation on a commercial scale; this is unlikely to take place before 1990. (For further reference to fast breeder reactors, see Section B).

The low and intermediate level wastes resulting from operation of large power reactors constitute less than 0.1% of the total radioactivity but account for a large volume. They are treated by conventional chemical engineering methods, and their treatment, containment and disposal operations can be carried out safely without any difficulty. These processes are developed in a high degree of efficiency and a decontamination factor of 100-1000 can be obtained depending on the initial activity and the treatment required to reduce the same to the stipulated levels. Particular care is always taken to remove specifically hazardous isotopes of elements such as strontium, cesium, radium, etc. Airborne wastes are passed through absolute filters which have efficiencies greater than 99.9% down to 0.3 micron particles (or 10^{-6} meters).

The record of radioactive waste management has been good so far and there is no reason to believe that it will not be maintained in the future. Current procedures explicitly involve exposures to workers and the public well below maximum permissible.

D. Use, Storage and Transport of Plutonium in a Nuclear Reactor Programme

In the course of the nuclear fuel cycle, there is a need to reduce as far as possible the number of steps involved with the handling of plutonium in order to improve and assist in its control. The proposal that all extracted plutonium should only leave any extraction plant in the form of complete fuel elements is one such approach. When plutonium is in a fuel element it is securely held and cannot escape unless the fuel is heated to a very high temperature (in excess of 1000 degrees Celsius). When plutonium has been extracted but is still in the extraction plant, it is considered to be in safe retrievable storage.

Safeguarding against the diversion of plutonium to atomic bomb manufacture requires very strict and careful accounting and ultimately some physical enforcement of security must be available. Such measures would involve continued national and international inspection and means for effective enforcement. These are certainly possible but require international cooperation and sacrifice of sovereignty.

It is likely to be forever impractical to take such a complete and accurate inventory of plutonium in extraction and processing plants that it would be possible to detect the abstraction of amounts sufficient to make several atomic bombs. Both direct measurements on separate process streams and less direct methods are therefore used, and will be further developed for detecting diversion to wastes or other purposes. Enforcement measures need further development. (See Section H below).

E. Transportation of Nuclear Materials

International regulations for the safe transport of nuclear materials drawn up by the International Atomic Energy Agency are available and satisfactory, and are revised periodically to take into account new developments. Improved techniques to implement these regulations will continue to be introduced. Risks of diversion during transportation differ widely in different locations and need to be met by enforced security measures, to appropriate security levels. Controls applicable to worldwide shipping are needed and call for strengthening of the Laws of the Sea.

F. Regional Reprocessing

The International Atomic Energy Agency is working actively on the question of regional fuel reprocessing facilities. First, a committee of experts from a number of member states will be preparing in the coming two years guides and standards for nuclear safety throughout the fuel cycle. A study of the soundness of the concept of regional nuclear fuel centres is included in the Agency programme. A special international conference on the whole nuclear fuel cycle is being planned by the Agency for 1977.

G. Training

The training of specialists for operating and safeguarding nuclear fuel production throughout its cycle will be an important and expanding need in the next several generations. International co-operation seems essential, and the International Atomic Energy Agency has begun a series of training courses, but will need increased financial support. A bottleneck could well develop in this area.

H. Enforcement

Even the best designed set of safety measures is useless unless they can be vigorously enforced. But such enforcement of security and safety measures presents a serious problem for which no satisfactory solutions have yet been found. International agreements with respect to such measures are needed urgently. It is to be hoped that the U.N. General Assembly, the Disarmament Conference and the International Atomic Energy Agency will give new attention to this problem. The churches might well make this one of their concerns.

I. Accidents

The question of the possibility of a large accident in a nuclear power station has been taken very seriously. Every large nuclear power reactor houses such enormous amounts of radioactive materials that under no circumstances should a sizeable fraction of these be allowed to get free. For this reason, considerable safety measures were introduced from the beginning of the atomic industry. Thus far any catastrophe has been avoided, though there have been incidents which might have become rather serious. These accidents demonstrate that no let up in vigilance must be allowed; rather a considerable increase will be required in view of the planned global mass deployment of large nuclear power plants. Growing familiarity with the constant danger, if it leads to carelessness or deviation from accepted safety procedures, will be a hazard in the years to come. The necessity of constant vigilance and the shortage of qualified personnel both impose a limit on the rate of growth of extensive nuclear power programmes.

J. Waste Heat

The waste heat burden for large nuclear power sites will be very considerable due to the volume of cooling water required and to the higher rejection of heat per kwh produced in the particular case of light water reactors. In both nuclear and fossil fuel plant used for electricity production, there will continue to be a limiting efficiency in the range of 36-40% so that the bulk of the heat produced is still likely to be rejected. Waste heat rejection to rivers, lakes and seas can cause severe problems in certain locations; dissipation through wet or dry cooling towers has potential related problems of localized meteorological changes. Constructive utilization of such waste heat including, for example, district heating and fish farming, is an immediate as well as a future requirement for more effective use of our energy resources. These waste heat problems will impose certain limits on siting of large nuclear power parks. Expected achievements and changes in nuclear power reactor designs and programmes, especially beyond the next decades, will assist in environmental adaptability to this stress.

V. NUCLEAR ENERGY AND NUCLEAR WEAPONS

The Hearing gave much attention to the question whether the extension of nuclear energy programmes will contribute to the further proliferation of nuclear weapons. Four critical issues were highlighted:

1. In most of the industrialized or technologically advanced countries, nuclear energy was developed on the basis of experience gained in the development of nuclear arms technology. In the present situation this "coupling" may tend to be reversed: the extension of the facilities for the production of nuclear energy to countries with developed nuclear science programmes may make it technically possible to construct nuclear weapons. The precise "tightness of the coupling", however, depends on the character of the nuclear programme and which atomic weapons are envisaged; it is a matter of vigorous debate! In the last analysis, this might not even be decisive, since a country, determined to obtain nuclear weapons, can do so by several routes. Nevertheless it is apparent that its efforts would be facilitated by the wide extension of nuclear energy technology.

2. It is difficult on political and moral grounds to deny countries without nuclear technology the right to obtain it because of a fear that they might use it for the development of nuclear weapons. The proposition that the appropriation of nuclear technology would forever be a limited right, to be doled out by the present nuclear countries according to rules determined by their interest is unacceptable. This would be an intolerable situation for many developing countries seeking to benefit from the peaceful application of nuclear energy and throw off technological domination by the already industrialized countries.

3. The major obstacle to nuclear disarmament is that the major industrial countries, already producing and possessing nuclear weapons, continue to regard them as indispensable for maintaining their power. Recent advances of military technology in the

U.S.A. and the U.S.S.R. may produce new instabilities in the strategic arms race between these two nations. The only convincing step toward preventing the nuclear arms race would be general nuclear disarmament by the present nuclear powers. This step would, however, become much more difficult if many yet non-nuclear countries were to obtain nuclear arms technology in the future, even if they do so on a very limited scale.

4. The Non-Proliferation Treaty is not a sufficient instrument to prevent nuclear arms' proliferation since it is based on discrimination in favour of countries already possessing nuclear weapons. It is not surprising therefore that it is difficult to enforce and achieve its universal acceptance and that many nations, especially several with a potential nuclear weapons capability, have so far refused to subscribe to it. It has not therefore fully prevented the spread of nuclear weapons (though Article VII of the Treaty encourages the formation of geographically extensive nuclear free zones).

It is part of the tragedy of our times that we see no simple way out of the present dilemma of our world: faced with the potentiality of nuclear energy and the fear of nuclear warfare. We are entirely agreed that nuclear weapons and nuclear warfare are a disastrous evil, and that by making nuclear arms nations are playing recklessly with the future of humanity. It is not clear, however, that stopping nuclear energy programmes will greatly reduce this hazard if countries are determined to equip themselves for nuclear war.

The churches and all others committed to a vision of justice and peace have a responsibility to sensitize people to these fundamental issues. Because of the fatalism about nuclear armament which is a characteristic of our times it is important to –

1. press for changes of social, political and economic structures which would help reduce international tension and conflict, and promote justice;

2. continue the struggle to persuade our countries to accept international control policies leading to effective disarmaments;

3 increase efforts to alert peoples to the enormous dangers, for all humanity and for future generations, of a nuclear war.

VI. NUCLEAR ENERGY
– A CHALLENGE TO THE CHURCHES

The decisions concerning nuclear energy depend on scientific, technical and economic judgements and choices; but also on religious and ethical assumptions. The churches are inevitably drawn into a discussion of the choices, not only by their own sense of responsibility but because scientists themselves are asking for public discussion and guidance on the issues involved. Every conversation during our meeting, no matter how technical the issue, soon became a discussion of human good, ethical decision and the meaning of human life.

We have found, to our surprise and pleasure, that, whatever our background and discipline, we have been able to speak and listen to each other with an openness of mind and spirit. We have not come to agreement on all points and in some instances our opinions still differ widely. But this meeting has for us all been a welcome experience in an area of discussion rife with misunderstanding, mistrust and polarization.

We call on all men of good will (and in particular all Christians), to commit themselves to this dialogue. All of us will in the coming years be expected to make critical decisions in this area and we shall need as much information and understanding of the issues involved as we can attain.

We call especially upon all churches to:

1. further understanding of the problems, both inside and outside the churches, making available to people at all levels of society, reliable and comprehensive information, and helping them to interpret and assimilate it;

2. provide occasions for dialogue where the tension between groups has grown so great as to bring conflict;

3. bring together those with special responsibility for policy in this field with other strata of persons, especially those they would not normally meet (e.g. problems arise where nuclear planners and engineers do not make an effort to meet with the population of the areas surrounding proposed nuclear power sites);

4. ask searching public questions on the goals of society and the means it uses, weighing long and short-term advantages and disadvantages, and in this light to provoke the community to examine critically its corporate and individual life styles;

5. point out that decisions taken in the field of energy ought to diminish and not increase global disparities in economic and political power.

SOCIAL AND ETHICAL ISSUES FOR REFLECTION AND ACTION IN THE CHURCHES

In continuing the dialogue we find the following points are important:

1. **Nuclear Energy and Social Goals.** Further discussion of the technicalities of nuclear energy should take place in a framework showing the relation to structural change and social justice, which implies a discussion of social goals and what constitutes progress. There is a tendency especially in the discussion of energy options to take as given the present emphasis on economic and energy growth in society. In this situation technological developments, like nuclear energy, tend to govern social ends so that social structures are inclined to conform to the dictates of technology. This problem should be one of the main foci of further ecumenical consideration of nuclear energy. This would be an especially fruitful theme for a continuing dialogue between natural scientists, technologists, theologians and social scientists, and political decision-makers.

2. **Nuclear Energy and the Distribution of Techno-Political Power.** The report raises the question of the impact of nuclear energy in increasing the disparity between the developed and the developing parts of the world. The concern for social justice and solidarity should lead us to press for a more equal distribution of techno-political power in the world. This issue requires immediate concern lest the expansion of nuclear energy increase these disparities. Moreover nuclear scientists tell us that international collaboration is the only realistic basis for a safe nuclear energy programme. What is entailed in the development of such a programme?

3. **The Ethics and the Method of Choosing Between Alternatives.** We have to make choices between different forms of energy supply. All alternatives must be studied and the same set of criteria applied to each before any choice is made. Yet in making the choice it is never possible to be absolutely sure about the right decision, e.g. it is impossible to eliminate all risks, or even to know exactly how great the risk may be. All human decisions are necessarily based on such imperfect and fallible knowledge. We are justified in demanding the most accurate possible knowledge of the advantages, risks and hazards of nuclear power, and also of the alternatives. Unique ethical problems may arise from the hypothetical nature of risk estimation with large-scale nuclear energy.

4. **The Limits of Technical Analysis.** While technical study can supply information on the risks and hazards of nuclear power, it cannot provide all the criteria by which these should be judged. A different kind of study is necessary to choose these additional criteria, and here we run into the difficulty that value systems vary between individuals and peoples. Criteria are needed to deal with such questions as:
 – the evaluation of risks to future generations;
 – the shape desired for society, or societies, in the future; (What values will shape those societies?)
 – how to distinguish needs from wants in a society?
 – the social structures of surveillance required by nuclear energy

systems, and their consequences at the political level. (Who finally decides and how?)

5. **Theological Critique of the World View Implied in the Nuclear Choice** (and by implication of the choice already made for a technological civilization). This world view involves an act of faith in humanity that it will be able intellectually, technically, psychologically, morally and spiritually, to master the awesome possibilities of nuclear power. The nuclear option implies a high level of discipline on the part of all people. It also requires a realistic assessment of the need for checks and balances necessary to minimize the effects of human folly and sin. On what assumptions is this faith credible? How far does it rely too much on human rationality, efficiency and achievement orientation?

6. **Rethinking the Faith Perspective.** The nuclear situation deepens the questioning of the presuppositions and goals of contemporary technological civilization, and the dilemmas posed by its achievements and its weaknesses become inescapable. It also brings us to such fundamental questions as: (i) How can men cope with their creativity? (ii) How can they live together in their diverse societies in a tolerable harmony? (iii) How far are they responsible for future generations? The Christian will want to seek help from his faith on these and other issues. A central message of the gospel is to respond to the coming of Jesus Christ by seeking God's kingdom and his righteousness, to live one's life in a joyful freedom from anxiety, not distraught by fears of the future, and looking with a sure hope that what God has begun in Christ he will complete in "a new heaven and a new earth". While in the gospel this confidence for living from day to day is set in a very limited human time perspective, in essence this confidence is the same now although our time perspective is immeasurably lengthened. But what precisely does this mean for human planning for the future? How far do recent scientific developments oblige us to reassess the whole scientific revolution from the 17th century out of which the harnessing of the powers of the universe which dominate our thoughts today has come? Is man "playing God", or is

he grasping creative techniques which God intends him to use responsibly? As humanity's scientific and technical capabilities increase, so do the stakes involved in its decisions become greater. The scale of success and failure becomes more dramatic. How far should the risks be run? How far would drawing back be a faithless failure of nerve? When is it necessary to say no in the name of God to something which is technically possible but which would be a betrayal of humanity to do?

All these different questions still have to be dealt with, in specific terms, and altogether, as part of the wider understanding of reality. Dialogue on such issues between Christians, with persons of other faiths, and with all persons of good will, has therefore to continue. The churches, and the World Council of Churches, are still just beginning to grasp the opportunity for probing the religious and ethical mysteries posed by a nuclear age. Attempting the challenge, in faith, they might still be a light to themselves and their world which is in deep spiritual confusion.

VII. SOME CONCLUSIONS

This Hearing has been a valuable experience to all participants. It has shown both the importance and the feasibility of a dialogue between people with widely differing backgrounds, coming from many different countries, but all of them seriously concerned about the problems nuclear energy may solve or create. It has given a wonderful opportunity to exchange views and to test, to widen or to correct personal opinions. We are pleased to share the results of our discussion with all those interested and involved in these questions. We hope it will offer some guidance to those who will have to make up their mind on these difficult issues. The rapid growth and increasing complexity of knowledge increases the need for this kind of interdisciplinary dialogue.

The Hearing has also left many of us in a state of perplexity, as we came to realize the meagreness of our knowledge and the momentousness of the decisions the world has to take. The presence among us of foremost experts in several branches of nuclear science and technology has not removed this uneasiness. No government, no organization, no individual can shirk responsibilities by referring simply to expert knowledge, for no one can be expert on all the physical, technical, geological, medical, biological, economical, social and political aspects of the nuclear issue. Weighing and synthesizing the opinions of many different experts and sifting the enormous amount of data available in published reports, tables, monographs and textbooks requires an "expertness" of a higher order. Therefore our group cannot put forward categorical recommendations. **It would not feel justified in either entirely rejecting, nor in whole-heartedly recommending large-scale use of nuclear energy.** All the same we can put forward some conclusions and outline some areas where our appraisals diverged.

1. We have been impressed by the reports on the safety of reactors. We are convinced of the deep sense of responsibility of many of those who design and operate nuclear reactors, and we are willing to concede that a nuclear power station operation in a stable world does not present greater risks than many chemical plants we have so far accepted without much misgivings. (Some of us may have even begun to worry more about such plants and about the vulnerability of our technological society in general than about nuclear reactors.) We have every reason to believe that at least comparable attention is being paid to the safety aspects of the fast breeder reactors which are being tested and which are potentially more dangerous. From this point of view there is no reason to draw the line between "burners" and "breeders"; there may, however, be other reasons for doing so.

2. Not all of us were aware of the fact that little reprocessing of spent nuclear fuel has as yet been taking place. The nuclear energy delivered thus far stems mostly from the fission of uranium 235; the plutonium fuel cycle is not yet in full swing. Even without plutonium extraction, nuclear reactors can produce nuclear power at comparable cost: this is shown by the Canadian example. But if we continue to work that way the world reserves of

cheap uranium will soon be exhausted. Nuclear reactors without fuel reprocessing can therefore provide only a temporary solution: they can help to bridge a gap until other energy sources become available; they do not offer prospects for a long lasting solution of the world's energy problems. Some people would be willing to accept nuclear energy, but only as a limited interim solution. They would therefore be against fast breeders, for fast breeders have been developed to use the plutonium recovered from the spent fuel. Those who would in principle be willing to accept a fuller nuclear programme, but who favour a very cautious progress will have felt relieved by the statement that breeders will not be commercial before 1990. This seems to assure that there will be much more research and testing before they come into widespread use.

3. Speaking of the hazards involved in a large-scale worldwide nuclear programme the American nuclear scientist Alvin Weinberg, has said: "The price we demand of society for this energy source is both a vigilance and a longevity of our social institutions that we are quite unaccustomed to." This statement was endorsed by most members of our group, but opinions differed widely as to the chances for such conditions being realized.

There are those who fear that the required vigilance will in the long run only be possible in an authoritarian society.

Those who are inclined to doubt the possibility of such vigilance and longevity, have either to reject nuclear energy entirely or to accept it only in a severely limited form, i.e. they might accept nuclear reactors without reprocessing and only in locations where favourable conditions for permanent storage of spent fuel exist. This would in no way obviate the need for vigilance with respect to wastes, but the dangers of plutonium theft at (or on the way to or from) reprocessing plants would disappear.

Others believe that the existence of large-scale nuclear energy will force countries to come together and that it will have a stabilizing and pacifying influence. Finally still others believe that the world is well-prepared for large-scale nuclear energy and that adequate security measures exist.

4. The position of developing nations received special attention. Some nations – e.g. India, Brazil, Argentina – have embarked upon extensive nuclear energy programmes. On the other hand, the economies of most of the African countries are not at present prepared to distribute and use the large quantities of electric power produced in a nuclear power station. For the time being more dispersed conventional power units of smaller capacity seem practical to meet the needs of the prevailing economy.

It was the opinion of many that nuclear energy will give the nuclear nations too much technological control over the developing countries as they accept technical assistance in the nuclear field; and the possible influence of large industrial firms on the nuclear economics and politics of such countries was viewed with alarm. (This situation might change as developing countries themselves become exporters of nuclear technology.) The Non-Proliferation Treaty was not acceptable to everyone, especially to those who considered that it interfered with the independence of a country; and some rejected the very idea of a treaty that recognizes the right of a few countries to make and possess nuclear weapons, while at the same time seeking to prevent other countries from doing the same.

There was a wide spread of opinions as to whether nuclear energy might possibly be of use to developing countries: some members were convinced that it will be an invaluable aid in fighting famine and misery, while others denied its potential usefulness altogether. All of us agreed that it will be of no avail without a more equitable world economic order.

5. It is regrettable but inevitable that one cannot speak about nuclear energy without thinking about nuclear weapons. For the nuclear nations, the former may be considered a spin-off of the latter; the armaments programme in these countries will hardly be influenced by the existence or non-existence of power producing reactors. For other countries the situation may be just the opposite: nuclear weapons will be a spin-off of nuclear energy. Certainly it is a political decision to make nuclear bombs, but it will

be much easier to carry out this decision if there already exists a general competence in nuclear technology.

So far the existence of nuclear weapons may have contributed to the avoidance of a Third World War. Will this continue to be the case when more and more countries possess nuclear weapons? All of us are deeply convinced that a peace based on nuclear deterrence is at best precarious.

It is not that kind of peace we are longing for, when we read in the gospel about peace on earth.

---

[1] Stock-taking of the recoverable reserves of coal, oil and natural gas will continue to depend for a number of years on the new levels of exploration activity in many countries, on economic factors (e.g. production costs) and on advances in technology.

[2] One isotope of uranium (U^{235}) is a nuclear fuel because it can undergo fission on impact with slow neutrons; each fission event yields an average of 2.5 additional neutrons making a self-sustaining chain reaction possible. Less than one percent of natural uranium is U^{235}; the remainder is nearly all U^{238} which is not readily fissionable.

[3] On absorption of a neutron the uranium isotope U^{238} undergoes changes leading to the formation of an isotope of the man-made element plutonium, i.e. Pu^{239}, which is fissionable under conditions similar to that of U^{235}.

[4] An otherwise "incredible" accident made up of a combination of very ordinary events; that is, outside the designer's normal frame of reference.

[5] The incorporation of radioactive waste material in molten silicates at high temperature resulting in the formation of a stable glass-like substance at normal temperatures.

[6] The REM (Roentgen Equivalent Man) is the unit of radiological dose applicable to absorption of ionising radiation in living tissue, one Rem corresponding to the absorption of one Rad of X or Gamma radiation (1 Rad = 100 ERGS per gramme).

[7] The Doppler effect is caused by the heating up of the atoms of uranium in the fuel thus causing them to move faster. Neutrons which are passing through the fuel tend to be captured by some of the U^{238} atoms at what is known as a "resonant energy". The increased velocity of the uranium

atoms increases the number of these atoms which are at the resonant capture energy relative to the passing neutrons. These U^{238} atoms therefore stop some of the neutrons which otherwise would have continued their travel until they were captured in the fission process, and this tends to reduce reactivity and power. An increase in fuel temperature will therefore be countered by a decrease in the power level on the reactor; the Doppler coefficient is therefore a negative power coefficient.

[8] The curie is the activity of a quantity of a radioactive isotope in which the number of atomic disintegrations per second is 3.70×10^{10}.

LIST OF PARTICIPANTS

I. Members of the Hearing Group

CASIMIR, Prof. H.B.G., Chairman (Netherlands)
    President, European Physical Society;
    President, Royal Dutch Academy for the Arts and Sciences

FRANCIS, Dr. John M., Rapporteur (U.K.)
    Senior Research Fellow in Energy Studies,
    Heriot-Watt University, Edinburgh

ALFVEN, Prof. Hannes (Sweden)
    The Royal Institute of Technology,
    Department of Plasma Physics, Stockholm

ANASTASSIADIS, Prof. Michael (Greece)
    Professor of Nuclear Physics, University of Athens

ARUNGU-OLENDE, Dr. Shem (Kenya)
    Electrical Engineer and Energy Specialist,
    U.N. Economic Commission for Africa,
    Division of Natural Resources, Science and Technology,
    Addis Ababa

BARNABY, Dr. Frank (U.K.)
    Director, Stockholm International Peace Research Institute

EHRENSTEIN, Prof. Dieter von (FRG)
    Professor of Nuclear Physics, University of Bremen

FARMER, Mr. F.R. (U.K.)
    Safety and Reliability Directorate,
    U.K. Atomic Energy Authority

FERGUSON, Dr. E.T. (Netherlands)
Research Physicist, Member of the Dutch Reflection Group
on Nuclear Energy

HAINES, Dr. John A. (U.K.)
Environment Directorate, O.E.C.D., Paris

HAMBRAEUS, Mrs Birgitta, M.P. (Sweden)
Member of Parliament

HEIDLAND, Prof. Dr. Hans-Wolfgang (FRG)
Bishop of the Evangelical Church of Baden

LEWIS, Prof. W. Bennett (Canada)
Department of Physics, Queen's University, Kingston

LUTZ, Dr. H.R. (Switzerland)
Superintendent, Nuclear Power Station

MACIEL, Dr. Lysaneas (Brazil)
President, Commission of Mines and Energy,
Camara dos Deputados, Brazilian Congress

NAGY, Dr. Mihaly (Hungary)
Institute of Nuclear Research,
Hungarian Academy of Sciences, Debrecen

NIFENECKER, Dr. Hervé (France)
Nuclear Physicist, Saclay – French Centre for Atomic
Research

OESER, Pfarrer Kurt (FRG)
Adviser of the EKiD on Environment Questions

POLLARD, Dr. William G. (U.S.A.)
Institute for Energy Analysis, Oak Ridge, Tenn.

RYDBERG, Prof. Jan (Sweden)
Chalmers Technical University,
Department of Nuclear Chemistry, Göteborg

SABATO, Dr. Jorge (Argentina)
Former Member of Atomic Energy Commission

SHEMILT, Dr. Leslie (Canada)
Dean of Engineering and Professor of Chemical
Engineering, McMaster University, Hamilton, Ontario

SHINN, Prof. Roger L. (U.S.A.)
Professor of Social Ethics, Union Theological Seminary,
New York

SIEGWALT, Prof. Gérard (France)
 Professor of Dogmatics, University of Strasbourg

SIMATUPANG, General T.B. (Indonesia)
 Member of the Supreme Advisory Council of the Republic
 of Indonesia

TERZIAN, Mr. Zaven-Bedros (Lebanon)
 Director, Centre for Information and Research on the Arab
 World, Beirut

THOMAS, Dr. K.T. (India)
 Director, Engineering Services Group,
 Bhabha Atomic Research Centre, Trombay

THUNBERG, Mrs. Anne-Marie (Sweden)
 Nordic Ecumenical Institute, Sigtuna

TIBELL, Dr. Gunnar (Sweden)
 Lecturer in Nuclear Physics, University of Uppsala

II. Technical Consultants and Advisers

MARCHETTI, Dr. C.
 Energy Project, IIASA, Austria

OFTEDAL, Prof. Per
 Geneticist and Radio-Biologist, University of Oslo

OPELZ, Mrs. Merle
 International Atomic Energy Agency, Geneva Office

OSHIMA, Prof. Keichi
 Nuclear Engineer and Director for Science,
 Technology and Industry, O.E.C.D., Paris

PRAWITZ, Dr. Jan
 Special Assistant for Disarmament
 to the Minister of Defence, Stockholm

III. W.C.C. Staff

ABRECHT, Dr. Paul
 Executive Secretary, Department on Church and Society

KOSHY, Mr. Ninan
 Churches' Commission on International Affairs

KROKER, Mr. Bruno
Senior Press Officer

SCHERHANS, Mr. Peter
Staff, Church and Society

STALSCHUS, Miss Christa
Staff, Church and Society

APPENDIX A

PRESENT SITUATION REGARDING THE PRODUCTION OF ELECTRICITY BY NUCLEAR POWER PLANTS

According to a recently published statement by staff of the International Atomic Energy Agency[*], the forecast of the world's nuclear electrical power in the period 1975-1985 is presented in Figure 1. The section of the curve up to 1980 represents the aggregate picture that has been assembled from governmental information on nuclear power plants that are currently in operation, under construction and planned for completion during this period. The growth of production capacity in the period 1980-1985 is obviously dependent on the availability of power plants which are now in the planning stage and for which a firm construction schedule is still to be finalized. The number of nuclear fuel cycle facilities to come into operation during the decade 1975-1985 is also presented in Table 1, and it is evident that in addition to a fourfold increase in the number of power reactors, there is virtually a doubling in the requirement for each stage of the fabrication, enrichment and reprocessing cycle in support of the expanded reactor programme. Since the specific reactor types are already known for most of these programmes, it is also possible to calculate the inventory of nuclear materials used for their operation. The divergence in the utilization of low enrichment uranium oxide based fuels as compared with natural uranium is indicative of the heavy commercial penetration of the world market by light-water reactor designs, similar to those adopted in the U.S.A. (Figure 2). The rate of plutonium production based on present operational experience with power reactors has also been calculated; the plutonium

[*] *Bulletin* International Atomic Energy Agency, Vol. 14, no. 2, April 1975, pp.4-14.

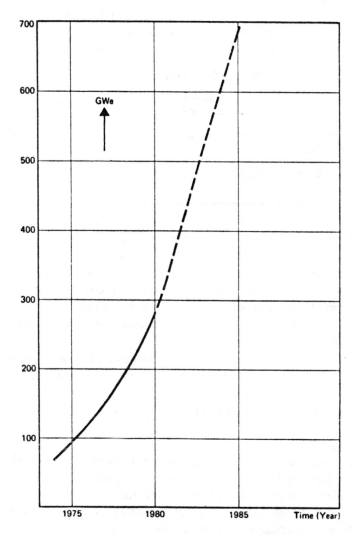

FIGURE 1

FORECAST OF THE WORLD'S NUCLEAR ELECTRICAL POWER
IN THE PERIOD 1975-85

TABLE 1
NUMBER OF FACILITIES IN THE NUCLEAR FUEL CYCLE

TYPE OF FACILITY	1975	1977	1980	1985
POWER REACTORS	200	280	420	800
URANIUM FUEL FABRICATION PLANTS	24	28	36	55
URANIUM AND PLUTONIUM MIXED OXIDE FUEL FABRICATION PLANTS	21	23	26	30
IRRADIATED FUEL REPROCESSING PLANTS	6	8	12	17
ENRICHMENT FACILITIES	8	9	10	13

inventory associated with the reactor programmes is shown in Figure 3 for both thermal and fast reactors and includes the cumulative amounts of plutonium in the spent fuel of thermal reactors. In some countries, the plutonium has not yet been extracted from the irradiated fuels but has simply been stored prior to the start of reprocessing; this is part of the Canadian policy with natural uranium fuels.

It is also worth describing the developing situation in terms of the positions that countries have adopted or are now adopting in relation to the prospect of expanding nuclear power programmes. At least these categories can be easily identified:

1. Countries which have nuclear power reactors in operation

These countries are listed in Table 2, which summarizes the position at the end of 1974. It can be seen that most countries have as yet made only a relatively small commitment to nuclear power generation, although there are substantial programmes of nuclear research under way in nearly all of these countries. The transition

FIGURE 2
INVENTORY OF NATURAL AND ENRICHED URANIUM
IN POWER REACTORS

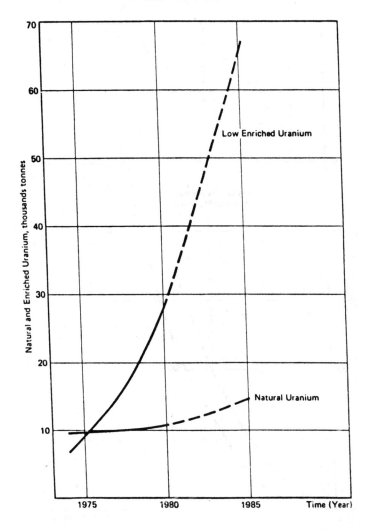

212

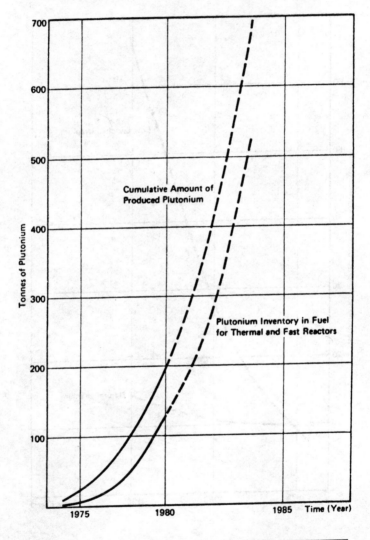

FIGURE 3

CUMULATIVE AMOUNT OF PRODUCED PLUTONIUM AND INVENTORY OF PLUTONIUM IN FUEL FOR THERMAL AND FAST REACTORS

TABLE 2
Countries which have nuclear power reactors in operation and have planned additions to installed capacity

Country	Number of Power Reactors*		
	1974	1975	1980
Argentina	1	1	2
Austria	–	1	1
Belgium	1	2	3
Brazil	–	–	1 (8)**
Bulgaria	1	2	4
Canada	7	6	12
Czechoslovakia	1	1	5
Finland	–	–	3
France	9	11	23 (1)
Germany, Dem. Rep.	2	3	3
Germany, Fed. Rep.	10	10	28 (4)
Hungary	–	–	1 (1)
India	4	3	8 (2)
Iran	–	–	– (4)**
Italy	3	3	7 (1)
Japan	10	11	29 (3)
Korea, South	–	–	2 (1)
Mexico	–	–	2
Netherlands	2	2	2
Pakistan	1	1	1
Spain	3	3	11 (2)
Sweden	4	5	11 (1)
Switzerland	3	3	8 (4)
Taiwan	–	–	4
Thailand	–	–	1 (1)
USSR	16	16	24
UK	31	31	39
USA	60	69	156 (37)
Total	169	184	391 (58)

Sources: 1974 and 1980 – The Nuclear Age, *SIPRI, 1974.*
  1975 – *"World List of Nuclear Reactors", in* Nuclear News, *18/10, p. 63, Aug. 1975.*
 * Output greater than 20 MW(e); the number in brackets after the 1980 figure is that of planned reactors not under construction on 1 January 1975
** Corrected according to latest newspaper reports.

that will take place during the period 1974-1980 is already quite dramatic as a significant number of countries, particularly in Western Europe, anticipate a threefold increase in the installed nuclear generating capacity.

2. Countries which plan to introduce nuclear power in the near future

Some countries have been included in Table 2 because they have definite plans to introduce nuclear power by the year 1980, viz. Austria, Brazil, Finland, Hungary, South Korea, Mexico, Taiwan, Thailand.

There are in addition other countries which plan to have nuclear power reactors in operation in the next decade, although firm commissioning dates have yet to be announced: Denmark, Iran, Israel, Philippines, Poland, Romania, South Africa, Yugoslavia.

3. Countries without plans for nuclear power plants

There are many countries, particularly in Africa, Latin America and Asia, which for technical and economic reasons have no plans to introduce nuclear power plants. These reasons include: small sizes of the power grid, sufficient supplies of other energy sources and lack of the required scientific and technical infrastructure.

APPENDIX B

NON-NUCLEAR ENERGY PROGRAMMES FOR DEVELOPING COUNTRIES

Electrical requirements of many developing countries in Africa, Asia and Latin America are still modest, and for technical and economic reasons will be met by small production units, for some time to come. Many of these countries have not reached the level of economic and industrial development to afford and support the introduction of nuclear energy installations. Neither do they have a nucleus of trained skilled personnel to help in the maintenance, operation and supervision of these installations. Furthermore, these countries lack the requisite organizational and institutional structures for the running of complex installations such as nuclear electricity generating installations including reprocessing plants and waste disposal systems or procedures. A large proportion of the population in these countries resides in the rural areas, where the consumption centres are relatively dispersed, and thus call for more decentralized supply systems.

It is in the interest of these countries to undertake a systematic survey and inventory of indigenous resources in order to enable them to plan the most suitable ways and means of meeting their energy requirements in the foreseeable future. In this context mention should be made of the prevalance of energy sources such as hydro, coal, natural gas, petroleum, oil shale, geothermal energy, solar energy, tidal and bio-conversion of plant and animal waste, etc.

Of these, bioconversion offers the greatest promise and should be given more serious attention. It can provide energy as well as fertilizers, although the deployment of this system of energy conversion may be limited by the availability of water resources. Solar energy is abundant in most of these areas, but it suffers from certain techno-economic as well as cultural limitations that must be overcome if its widespread utilization is to become a reality.

Better conservation procedures as well as more efficient and effective use of existing installed electricity generating capacity in these countries is also called for. In addition, it is in the interest of many of them to cooperate in the joint development of hydro-electricity in order to share the high initial capital costs and ensure an adequate market; in the search for new deposits of coal, oil and natural gas; in adopting measures (with oil producing countries) that would guarantee the supply of petroleum and its products under commonly acceptable conditions, etc.

In the longer run, there may be scope for the introduction of nuclear plants into the power systems of these countries, when a sufficiently large number of them will have systems large enough to absorb nuclear plants. The countries will also be on a stronger industrial and economic footing which will enable them to support a nuclear electricity generating installation and have the technical and organizational capacity to maintain, operate, supervise and manage such complex systems.

There are, of course, countries in Africa which already have large enough power systems and an adequate economic and industrial base to support a nuclear energy installation including fuel processing plants. It is hoped that in their deliberations and decision on the nuclear option these countries will take into account the availability and supply of other sources and forms of energy, the costs and benefits of these and the safety requirements and environmental and social-economic consequences of whatever option – nuclear or non-nuclear – is decided upon.

APPENDIX C

SUMMARY OF RESEARCH AND DEVELOPMENT OF ENERGY RESOURCES

The position concerning future prospects has been ably summarized in a recent O.E.C.D. report (Energy R / D: Problems and Perspectives, Paris 1975); the main results relating to the more promising and realistic solutions are as follows:

The short-term (1974-1985)

In the short term, energy needs will be satisfied mainly by the same resources and technologies as exist today. Nevertheless, R

& D activities, and more particularly those which already exist, can make significant contributions by discovering new resources and by improving existing technologies.

(i) **Oil and natural gas** will continue to play a major role. Among the most important possible contributions of R & D are:

- the improvement or prospection technologies on and off-shore;
- the improvement of secondary and tertiary recovery methods;
- better deep-drilling technologies;
- with regard to off-shore technologies, it should be possible to produce oil in increasingly deep waters, up to an ocean depth of approx. 1,000 m.

(ii) **The direct use of coal for electricity generation** will probably become more extensive. The most relevant contribution of R & D in this sector will be the development of new stack-gas cleaning technologies.

(iii) **Nuclear energy for electricity generation** will be developed very rapidly. This development will essentially be based on already existing reactor types; mainly light water reactors and to a much smaller extent, heavy water reactors. In the nuclear energy sector, the greatest part of the R & D effort will have to focus on the fuel cycle.

Several energy sources could make contributions which, though limited in overall terms, might be quite significant in specific cases, for example:

- shale-oil in-situ extraction in the United States;
- tar-sand oil extraction in Canada;
- geothermal energy through dry steam;
- solar energy for space heating and cooling;
- synthetic fuels through pyrolysis of organic waste.

As to **energy carriers,** it is probable that electricity technologies will develop most rapidly. R & D will continuously improve the efficiency of electricity generation and reduce thermal pollution problems, through cooling towers, increasing utilization of the residual heat of power plants, etc.

Finally, the analysis of **energy systems** should be pursued with greater speed and scientific effort. The contribution of systems analysis to a rational and efficient solution of energy problems could certainly make itself felt in the short term, but it is also important for the medium and longer term.

The medium-term (1985-2000)

R & D now being pursued will have a much broader impact in the medium than in the short term. With regard to energy resources, it should be possible to produce **oil** from an **ocean depth** of considerably more than 1,000 m, and **tertiary recovery methods** will probably be applied on a broad, general scale. Better **coal conversion** (gasification, liquifaction, conversion into methanol), and advanced **coal firing** technologies will confirm the importance of coal for electricity generation and for synthetic fuel production. In addition, **shale-oil** extraction could spread to many countries which hitherto have made no systematic effort to discover and assess shale resources. Finally, **nuclear energy** will occupy an increasingly important place, following the introduction of **fast-breeders** and of **high temperature gas cooled reactors.**

Again, certain **non-conventional energy sources** could play a globally limited but locally very important role. Among these sources are geothermal energy through hot brines, solar energy for electricity generation, synthetic fuels through bio-conversion of plants or organic waste, wind-power, etc.

Electricity will probably remain the energy carrier with the fastest development potential. Electricity generation could benefit from progress in advanced cycles and perhaps fuel cell technologies, electricity storage from progress such as that made with batteries and electricity transmission from that with underground cables. Moreover, electricity will profit from increasing possibilities of application such as electric cars. Other energy carriers will become increasingly important, for example methanol or others which can store and transport heat in the form of chemical bond energy.

It is in the medium term that research to improve **energy utilization** will be most successful. A more efficient energy utilization at the level of individual technologies might help considerably to slow down the growth rate of energy demand.

The long-term (after 2000)

The long-term prospects are obviously much more uncertain. **Fossil fuels** will continue to play a noteworthy role, especially **coal**, as coal reserves are known to be sufficient for several centuries at the present rate of consumption. However, progressively, fossil fuels will have to give way to energy sources which have, if not "unlimited", then at least extremely abundant reserves. Thus, in the very long term, fast breeders, controlled thermonuclear fusion, solar energy and last, perhaps geothermal energy through hot rocks will increasingly replace present energy sources.

As far as **energy conservation** is concerned, the biggest long-term impacts of R & D will among others, be found in the transportation and agricultural sectors.

APPENDIX D

SWEDISH OPINION POLLS ON NUCLEAR ENERGY

In Sweden after a nation-wide information campaign supported by government money, several polls were made during the spring. Some of these were carried out within the five political parties with representation in the Swedish parliament. In addition to a choice between various alternatives regarding nuclear energy the persons interviewed were asked about their opinion on the energy consumption in general. As an example of the results obtained in these polls we quote the following ones which date from March of this year. They pertain to people who voted for the social democrats (the government party) in the last election.

No nuclear energy at all	26%
Nuclear energy as a last resort	37%
Nuclear energy is necessary for keeping our standard of living	18%
Uncertain	19%
Total	100%

On the question of the energy consumption in general the following answers were given:

Energy consumption should be kept constant	37%
Energy consumption could decrease	36%
Moderate increase recommended	19%

As a final remark it may be mentioned that a decision in principle has been made (as of June 1975) to build 13 reactors in Sweden on four different sites totalling an effect of approx. 10,000 MW(e) which would imply a fraction of about 10% of the total energy production coming from nuclear power plants. Of the total electric power 30-40% would be nuclear when the programme is completed (around 1985). The nuclear energy programme will be reviewed again in 1978, using the experience gained in the meantime both within Sweden and abroad.

SECTION 6
A STEP FORWARD

I. RESPONSES TO THE SIGTUNA REPORT

The "Report on Nuclear Energy" produced by the programme Sub-Unit on Church and Society of the World Council of Churches has already served a valuable purpose in many countries by stimulating an immediate level of concern over the future implications of the growth of nuclear power generation. The report was originally published in a limited edition of the journal *Anticipation* (No. 21, October 1975); the copies were so quickly in demand that it was decided to make the document more widely available and that is the purpose of this edition.

The material has been of particular interest to nuclear scientists, government policy makers, industrial and political leaders, as well as theologians and social ethicists, and has been generally welcomed especially in countries where the debate about the extension or the introduction of nuclear energy programmes is under way. This includes Japan, New Zealand, Argentina, Brazil, Holland, the U.K., West Germany, France, Switzerland and the U.S.A.

There have been many favourable reactions to the report both from those opposed to as well as those in favour of nuclear energy, but quite apart from the two main poles of the argument it may be of interest to the general reader to see the nature of the response that has so far been provoked.

Dr. Magnus Pyke, Secretary of the British Association for the Advancement of Science, writes:

"The report on nuclear energy... is absolutely first class ...it contains a wealth of valuable material and sets out admirably the ethical principles underlying the whole problem."

Writing in *The Times* (London) for December 15, 1975, Bishop Hugh Montefiore says:

"The public hearings held at Sigtuna in Sweden earlier this year under the auspices of the World Council of Churches... provide a splendid example of what is urgently needed in this country (U.K.) now before decisions are made which may lead to 'inevitable' programmes."

From New Zealand, Sir Guy Powles, Director of the Office of the Ombudsman, writes:

"The issue of *Anticipation* on Nuclear Energy is... a really magnificent and most timely document... The issue presented is one of the greatest of our time, and the thoughts and presentations put forward are of great value, particularly in countries like mine where we seem to be trembling on the edge of the nuclear brink in more ways than one."

Dr. David Rose, Professor of Nuclear Engineering at M.I.T., in a paper addressed to legislators and scientists concerned with energy, notes "that the W.C.C. study is the best general assessment that I have yet seen" and "is of outstanding quality."

Some commentators, particularly in the U.S.A. where the nuclear energy debate is already polarized, have found the W.C.C. report "equivocal" and Professor Charles West, commenting on the debate among scientists at Princeton University, notes:

"The statement produced by the ecumenical hearing is known, but tends to be ignored by the pro-nuclear energy forces and vigorously opposed by the anti-group. Such is... the fate of reason and moderation when political conflict is the name of the game."

In response to this last quotation, it is perhaps necessary to add that the almost unique circumstance of the ecumenical hearing was the direct engagement of the many different factions already committed in the nuclear debate. Not only would it have been impossible for such a varied group to produce anything approaching a consensus statement on such a controversial subject, but the struggle to achieve such a statement would have totally denied the integrity of each individual expressed through his or her contribution to the hearing process. There was undoubtedly common ground that could be shared by the majority of the participants and this is clearly reflected in the body of the report. At the same time it must be acknowledged that there were major differences of opinion over the social responsibilities of scientists, engineers and others whose task it is to secure the nuclear fuel cycle in every sense.

In certain respects, the events of that week in Sigtuna were the preliminaries in a continuing act of reconciliation that the World Council of Churches should now engage at the level of the international community. The problems surrounding the large scale development of nuclear energy demand widespread scrutiny *before* they reach unmanageable proportions.

During the past year (1975–76) a group of seven nuclear exporting nations – Canada, France, Japan, U.K., U.S.A., U.S.S.R. and West Germany – has met in secret to define stringent new conditions for securing the so-called "sensitive technologies" associated with management of the nuclear fuel cycle. These sensitive technologies are (1) the enrichment of uranium, (2) the reprocessing of spent nuclear fuel (*N.B. both of these processes yield fissile material from which nuclear weapons can be made*) and (3) the refining of heavy water, which is a key requirement for the operation of certain types of nuclear reactor. The new terms for safeguarding exports of these three technologies were made public in a parliamentary reply by the U.K. Foreign Secretary on 31 March 1976:

– *First*, the exporting nation will require an assurance that nuclear technology of any description will not be used to manufacture a nuclear explosive for any purpose, including so-called peaceful nuclear explosives (PNEs) designed specifically to perform an earthmoving task. This stipulation was not made before in contracts negotiated with nations which had refused to sign the Non-Proliferation Treaty (NPT).

- *Secondly*, the exporting nation will require an assurance that exports will be adequately safeguarded against the risk of theft or sabotage. This is an entirely novel requirement, and reflects the growing public concern with the activities of terrorists and dissident political groups.

- The *third* requirement concerns the re-exporting of nuclear technology transferred from one nation to another. The recipient will be required to give assurances that the technology will be re-exported only under the same international safeguards that covered the initial transfer of the technology.

- The *fourth* requirement covers "replication", the possibility that technology transferred from one nuclear facility under international safeguards might then be used by the recipient to construct other facilities beyond the reach of those safeguards. The supplier will require assurances that safeguards will apply to any replication of plants within 20 years.

While all seven member nations of the nuclear suppliers' group agreed on the same assurances, another three nations – The Netherlands, Sweden and East Germany – have joined the group. It is expected that in addition Belgium, Italy and Czechoslovakia will also comply. With these negotiations still in the very early stages, it is clear that the present agreement cannot be considered as the ultimate in safeguards. The United States, in particular, is still pursuing its suggestion of regional fuel cycle plants operated under international control to prevent misuse of enrichment or the plutonium by-product.

It is vitally important that non-governmental organizations, such as the World Council of Churches, should be prepared to see these problems on a very much longer timescale than that conceded by the national governments with a direct interest in utilizing nuclear technology for a variety of motives. The implications of this technology distributed on a world scale must now be clearly identified for the sake of all humanity. That could be part of an ongoing task for the churches at a national and at a community level.

II. REACTIONS FROM CHURCHES

In several countries churches have begun to respond to the social and ethical challenge posed by nuclear energy. They also want to encourage a wide-spread public debate. The W.C.C. Report is being used as a starting point for discussion and education in Scotland, Denmark, New Zealand and Argentina. For example, a summary of the W.C.C. Report was submitted to the Assembly of the Church of Scotland with a proposal urging "the British Council of Churches to convene an ecumenical hearing in this country (U.K.) on the scientific, technical and moral issues of the expansion of nuclear power, especially on the introduction of the fast breeder reactor".[1] In Denmark and New Zealand the churches are making plans to encourage and to participate in a developing public debate about the policy of introducing nuclear energy into these countries.

In other countries where the debate about existing nuclear power policies is already underway use has been made of the Sigtuna Report, sometimes favourably, sometimes critically, to look at the social and ethical issues. This is true in Holland, Switzerland, West Germany and France. For example, in Germany a group of theologians and scientists have formulated a statement on the *Need for a Moratorium*, questioning the Federal Government's plans for increasing development of nuclear power and outlining a 9-point programme:

"1. We demand in the first place a drastic reduction and slow-down of existing plans for nuclear power plants.

2. We advocate intensification of research efforts into the controversial issue of risk.

3. We demand a medium-term programme of energy conservation which, besides covering energy needs would provide time and scope for the development of new energy production technologies. To avoid any misunderstanding: energy conservation through better use of the energy produced, even without any increase in the primary energy used, does not mean a halt to economic growth.

4. We consider that a reduction of the annual rate of increase of primary energy input is urgently needed.

5. We advocate an immediate increase of technological development of new sources of energy (e.g. solar energy, nuclear fusion). These projects cannot conceivably be implemented without international cooperation. We stress the importance of traditional energy sources such as coal, petroleum and natural gas for the short- and middle-term energy supply in West Germany. For the long-term, neither fossil nor nuclear fuels are ideal energy production processes.

6. We underline the need to take ecological and regional planning criteria into account in the development of energy production at both national and international level. In West Germany we are in the process of destroying the ecological balance by subjecting the last remaining agricultural areas to an expansionist energy and industrial policy with serious consequences for the environment.

7. We demand a clear definition of priorities in the energy policy of West Germany, responsibly controlled by the proper democratic channels and with all due attention to energy conservation measures, protection of the environment, alternative sources of energy production and appropriate safeguards in the foreign sector of the economy.

8. We warn against one-sided and uncontrollable determination of scientific and technological policy such as has led to the choice in favour of nuclear energy. We support all measures

undertaken by the Federal Government leading to greater openness in political decision-making regarding technological innovations in the energy sector. This implies, among other things, that the area between different prognoses and scientific assessments will be the subject of thorough and public discussion.

9. We criticize the mistaken and widely propagated idea of national self-sufficiency for West Germany in the matter of energy provision and stress the security safeguards inherent in international economic involvement and interdependence. The economic future of West Germany and the stability of her social conditions can only be guaranteed by international economic cooperation and a diversified energy programme. Furthermore, the possibility cannot be excluded that the uranium-supplying countries will band together to form an OPEC-type organization."

The W.C.C. Report on Nuclear Energy has received particular attention in the U.S.A. where its publication coincided with a statement prepared by a Committee of Inquiry on the Plutonium Economy, arranged in cooperation with the Division on Church and Society of the National Council of Churches.

The statement of the Committee of Inquiry adopted a strong position against the plutonium economy and against the expansion of nuclear power, and called for a moratorium on development of a commercial breeder reactor.

This position was criticized by the nuclear industry and many scientists. A meeting to hear the critics was organized by the NCC in New York City on January 28, 1976. Later the Statement on the Plutonium Economy was withdrawn and replaced with a resolution on the Plutonium Economy asking for the "Government of the United States to declare a moratorium on the development of a plutonium economy until the people of the United States can assess the pros and cons of the issues, the said moratorium to be defined as follows:

A moratorium on the commercial processing and use of

plutonium as an energy source, and on the building of a demonstration plutonium breeder reactor, pending further study of the theological, economic, socio-political and technical issues involved."

This was adopted by the Governing Board of the NCC at its meeting in Atlanta, Georgia, March 2–4, 1976. In the report to the Governing Board of the National Council of Churches U.S.A., the NCC Division of Church and Society compares its own approach to the problem of "the Plutonium Economy" with that of the W.C.C. Hearing on the Problems of Nuclear Energy, as follows:

"The goal of the Division of Church and Society was to raise the issue of plutonium use for full and wide discussion within the churches, an end which could best be accomplished by suggesting a strong position on one side or the other of the plutonium debate. In contrast, the World Council of Churches was simultaneously studying the wider field of nuclear energy, using a process of gathering opinions from people on all sides of the issue. The predictable result was a report which stated both the pro and con arguments, stressed the importance of the issue, but neither rejected nor supported nuclear energy use.

"While we respect the World Council's choice of process, it was the Division's conviction that, since people are being asked to make decisions for or against nuclear energy today, to remain 'neutral' is in practice to support continued growth of nuclear capacity in the U.S. and increasing reliance upon it. The World Council's 'balanced' report failed to arouse debate on the issues within the U.S. Church community.

"In any case, the NCC had chosen to concentrate on the use of plutonium as a fuel, rather than on the wider use of nuclear energy, so comparison of the two processes is not very meaningful. The current generation of nuclear reactors uses uranium as a fuel; the spent fuel contains plutonium. It is the wisdom of reprocessing this plutonium for use in current reactors, and pursuing development of plutonium breeder-reactors, that was questioned in the Committee of Inquiry Report and the policy statement proposed to the NCC in October 1975."

The debate in the U.S. has now moved forward into an even more critical phase with the drafting of proposition 15 on the California primary ballot paper.

In June 1976, the people of California were asked to cast their votes on the Californian Nuclear Safeguards Act. The initiative took the following form:

—To prohibit nuclear power plant construction and to prohibit operation of existing plants at more than 60% of
the original licensed core power level unless federal liabil-
ity limits were removed.

—After five years—to require derating of plants 10%
annually unless the state legislature, by two-thirds vote,
confirmed the effectiveness of safety systems and waste
disposal methods.

—To permit small-scale medical or experimental nuclear
reactors.

—To appropriate $800,000 for the expenses of a fifteen
person advisory group and for legislative hearings.

If the proposed initiative had been adopted, this would have considerably restricted the operation of existing nuclear power plants. In the event the ballot showed a substantial majority against the adoption of the initiative but undoubtedly this will create an important precedent for a number of other countries to follow the experience of both Sweden and the United States; it is the cumulative effect of this process of public consultation which cannot be anticipated. In a number of industrialized countries, decisions on the planning of nuclear power programmes are not subject to public scrutiny but are instead regarded as strategic decisions taken "in the national interest". However, it is clear that until there is a much more open public debate on the subject of large-scale nuclear energy programmes, decisions will continue to be made in the more restricted context of official government agencies.

III. COMMENTS AND REACTIONS OF THE NAIROBI ASSEMBLY

The Fifth Assembly of the World Council of Churches meeting in Nairboi, Kenya, November 23 to December 10, 1975, received the report on Nuclear Energy and in turn transmitted the following statements to the churches:

Nuclear Energy. Increasing pressures on the world petroleum market bring the prospects of rapid expansion in the construction of nuclear power stations in countries at widely differing stages of development. This poses ethical dilemmas in view of (1) the inevitable coupling of nuclear energy for electricity generation and the production of nuclear weapons: how can we limit the use of nuclear energy to the peaceful purposes of development, and avoid the possibility of using nuclear energy in nuclear weapons? (2) the hazards involved in the storage of nuclear waste for long periods; and (3) the problems of theft and sabotage. The ecumenical discussion of these issues has just begun and needs to be continued. Discussion should concentrate on the effectiveness of the monitoring and control of the nuclear fuel cycle at the international level, ratification of the Non-Proliferation Treaty by all countries possessing nuclear technology, the drift towards a plutonium fuel economy and its consequences, and the political and military implications of the further spread of nuclear technology.

Alternative Energy Technologies. During the next decade, there will be increased research and development of alternative systems of energy production. The world is now entering the transition between dependence upon petroleum to dependence upon other sources of energy, e.g. solar, wind, water and geo-thermal. Priority must be given to ensure that research on alternative energy sources is funded at a high level, especially as these technologies will be of direct benefit to developing countries.

These statements were then supported by the following recommendations to the churches:

Nuclear Energy. As far as nuclear energy is concerned the churches at the national level should lobby for more effective safeguards in the control and operation of the nuclear fuel cycle to ensure that there is a reduced risk of diversion of strategic weapon materials; in addition there is a need to press for more serious investigation of the long-term disposal of radio-active wastes by those responsible for the management of nuclear power programmes. The churches are further urged to register their reactions to the findings of the Ecumenical Hearing on Nuclear Energy, Sigtuna, Sweden, 1975.

Alternative Energy Technologies. Since there are inevitable grave ethical problems involved in the application of nuclear energy, we recommend that the churches and individual Christians seek to promote sufficient funding and moral support for research and development of renewable sources of energy, e.g. solar and wave energy.

IV. THE ISSUE OF A MORATORIUM

In the present situation of uncertainty concerning the potentialities of nuclear energy for good or for evil, several countries are practising or at least considering a moratorium on nuclear developments and some churches will seek to exert an influence on the public in that direction. A moratorium might be directed at the total production of nuclear energy or at only certain specified reactor types (e.g., the fast breeder reactor). It might be aimed at deferring the commercial production of electrical energy from a particular system or it might be extended to include experimental projects.

Obviously a moratorium is only a delay. It defers the decision to proceed with nuclear production or to stop it. Sooner or later the decision must be made. Determined advocates or opponents of nuclear energy will only favour a moratorium for tactical reasons. Many who favour the introduction of a moratorium put forward two substantive arguments: (1) A moratorium may slow the momentum of a process that some think is moving too fast. The delay may give time for further public discussion, so that the public is not confronted with a *fait accompli*. (2) A moratorium may give the opportunity for specific research on technical points where more information is needed for decisions.

Those who oppose the principle of a moratorium use two different arguments: (1) A moratorium on nuclear energy, unless combined with radical social changes not now in process, increases the reliance of industrialized nations on scarce energy sources, thereby reducing the energy available to poorer countries. Since the industrialized countries can pay for such energy (e.g., by importing petroleum), the hardship falls on countries that cannot compete financially. There is some evidence that if country A reduces expectations of nuclear energy production and thereby increases its oil imports, country B feels pressure to speed up its nuclear production in the face of scarcer petroleum supplies. (2) In the case of experimental programmes with long lead times, a moratorium, if really effective, may produce an energy crisis of a different kind at a later date.

Whatever the standpoint of the individual reader, personal advocacy of a moratorium is not simply the delay of a decision or an easy way to avoid a decision. It is itself a decision with serious social consequences.

V. NUCLEAR ENERGY AND GLOBAL SOCIAL JUSTICE

Will the spread of nuclear energy contribute to or weaken the struggle for social justice within and between nations? Some oppose nuclear energy because they believe that for the forseeable future it is of the greatest benefit only to the rich and to the technologically developed nations. Moreover, in the name of "security" the nuclear nations tend to band together to guard the sharing out of the sensitive nuclear technologies, thus producing the secretive nuclear club, determining which nations will have access to nuclear energy and defining the appropriate constraints. The sale of nuclear technology to countries with repressive social and racial policies has further heightened the fear of many that nuclear energy will increase injustice in the world. And many in the poorer countries are convinced that the political criteria used to determine the sharing of nuclear technologies serve only the interests of the already powerful nations.

Many also believe that sophisticated nuclear technology has unfavourable social implications, accentuating the formation of industrial-urban concentrations and the centralization of technical power which some countries are now beginning to query. Nuclear energy seems thus to work against the movement for smaller scale technology and directly against "alternative", perhaps more socially appropriate, energy systems.

However, to many developed and developing nations nuclear energy appears to offer freedom from dependence on other foreign energy resources and they assert their right to benefit from nuclear power development. Many are in a position (e.g., Cuba, Brazil, Jamaica) where nuclear energy could free them from dependence on expensive imports of oil or coal. It is significant for example, that at the recent U.N. Habitat Conference in Vancouver only one of the developing countries (Papua New Guinea) opposed nuclear energy.

If it is true that the world of today has no viable alternative other than to extend the nuclear option to more and more countries, then it becomes all the more important to examine the social implications of nuclear energy and to determine the social and political conditions under which it may be used. These conditions must necessarily express the concerns of all nations and not just the informed fears of those already possessing nuclear capability.

The question governing the world of tomorrow may continue to be the one we have in mind today, that is, "Who guards the guardians?" (*Quis custodiet ipsos custodes?*). In all humility and in a spirit of true repentance, we cannot escape the uncertainty of our response.

> "Who sees with equal eye, as God of all,
> A hero perish, or a sparrow fall,
> Atoms or systems into ruin hurled,
> And now a bubble burst, and now a world."
>> *Alexander Pope, An Essay on Man*

[1] Submission on Nuclear Power, International Sub-Committee, Church and Nation Committee, Church of Scotland General Assembly, May 1976.

GLOSSARY OF TECHNICAL TERMS

Alpha particle – A positively charged particle emitted by certain radioactive materials. It is made up of two neutrons and two protons bound together, and is identical to the nucleus of a helium atom.

Atom – A particle of matter indivisible by chemical means – the fundamental building block of the chemical elements.

Atomic bomb – A bomb whose energy comes from the fission of heavy elements, such as uranium or plutonium.

Atomic number – The number of protons in the nucleus of an atom – also its positive charge.

Boiling-water reactor – A reactor in which water, used as both coolant and moderator, is allowed to boil in the core. The resulting steam can be used directly to drive a turbine.

Breeder reactor – A reactor that produces more fissile (fissionable) fuel than it consumes. The new fissile material is created by capture in fertile materials of neutrons from fission, a process known as breeding.

Burner reactor – A reactor that produces some fissile material, but less than it consumes.

Chain reaction – A reaction that stimulates its own repetition. In a fission chain reaction a fissionable nucleus absorbs a neutron and fissions, releasing additional neutrons. These in turn can be absorbed by other fissionable nuclei, releasing still more neutrons. A fission chain reaction is self-sustaining when the number of neutrons released in a given time equals or exceeds the number of neutrons lost by absorption in non-fissioning material or by escape from the system.

Containment – The provision of a gas-tight shell or other enclosure around a reactor to confine fission products that otherwise might be released to the atmosphere in the event of an accident.

Control rod – A rod, plate, or tube containing a material that readily absorbs neutrons, used to control the power of a nuclear reactor. By absorbing neutrons, a control rod prevents the neutrons from causing further fission.

Conversion ratio – The ratio of the number of atoms of new fissile material produced in a burner reactor to the original number of atoms of fissile fuel consumed.

Coolant – A substance circulated through a nuclear reactor to remove or transfer heat. Common coolants are water, air, carbon dioxide and liquid sodium.

Core – The central portion of a nuclear reactor containing the fuel elements and usually the moderator, but not the reflector.

Critical mass – The smallest mass of fissile material that will support a self-sustaining chain reaction under stated conditions.

Criticality – The state of a nuclear reactor when it is sustaining a chain reaction.

Curie – The basic unit to describe the intensity of radioactivity in a sample of material. The curie is equal to 37 billion disintegrations per second, which is approximately the rate of decay of one gram of radium.

Decay, radioactive – The gradual decrease in radioactivity of a radioactive substance due to nuclear disintegration. What remains is a different element. Also called radioactive disintegration.

Depleted uranium – Uranium having a smaller percentage of ^{235}U than the 0.7 per cent found in natural uranium. It is obtained from the spent (used) fuel elements.

Doubling time – The time required for a breeder reactor to produce as much fissile material as the amount usually contained in its core plus the amount in its fuel cycle (fabrication, reprocessing, and so on). It is estimated as 10-20 years in typical reactors.

Electron – An elementary particle with a unit negative electrical charge and a mass 1/1836 that of the proton.

Element – One of the 105 known simple substances that cannot be divided into simpler substances by chemical means. A substance whose atoms all have the same atomic number.

Enriched material – Material in which the percentage of a given isotope present in a material has been artificially increased, so that it is higher than the percentage of that isotope naturally found in the material. Enriched uranium contains more of the fissionable isotope ^{235}U than the naturally occuring percentage (0.7 per cent).

Enrichment – A process by which the relative abundances of the isotopes of a given element are altered, thus producing a form of the element which has been enriched in one particular isotope.

Fallout – Airborne particles and other matter containing radioactive material which fall to the ground following a nuclear explosion.

Fast breeder reactor – A reactor that operates with fast neutrons and produces more fissile material than it consumes.

Fast reactor – A reactor in which the fission chain reaction is sustained primarily by fast neutrons rather than by slow (thermal) or intermediate neutrons. Fast reactors contain little or no moderator to slow down the neutrons from the speeds at which they are ejected from fissioning nuclei.

Fertile material – A material, not itself fissionable by thermal neutrons, which can be converted into a fissile material by irradiation in a reactor. There are two basic fertile materials,

^{238}U and ^{232}Th. When these fertile materials capture neutrons, they are partially converted into fissile ^{239}Pu and ^{233}U respectively.

Fissile material – While sometimes used as a synonym for fissionable material, this term has also acquired a more restricted meaning, namely, any material fissionable by neutrons of all energies, including (and especially) thermal (slow) neutrons as well as fast neutrons: for example ^{235}U and ^{239}Pu.

Fission – The splitting of a heavy nucleus into two approximately equal parts (which are nuclei of lighter elements), accompanied by the release of a relatively large amount of energy and generally one or more neutrons. Fission can occur spontaneously, but usually is caused by nuclear absorption of neutrons or other particles.

Fission products – The nuclei (fission fragment) formed by the fission of heavy elements plus the nuclides formed by the fission fragments' radioactive decay.

Fuel – Fissile material used or usable to produce energy in a reactor. Also applied to a mixture, such as natural uranium, in which only part of the atoms are readily fissionable, if the mixture can be made to sustain a chain reaction.

Fuel cycle – The series of steps involved in supplying fuel for nuclear-power reactors. It includes mining, refining, the original fabrication of fuel elements, their use in a reactor, chemical processing to recover the fissionable material remaining in the spent fuel, reenrichment of the fuel material, and refabrication into new fuel elements.

Fuel element – A rod, tube, plate, or other mechanical shape or form into which nuclear fuel is fabricated for use in a reactor.

Fuel inventory – The amount of fissile material in the core of a reactor.

Fuel reprocessing – The processing of reactor fuel to recover the unused fissile material.

Fusion – The formation of a heavier nucleus from two lighter ones (such as hydrogen isotopes), with the attendant release of energy (as in a hydrogen bomb).

GWe – One thousand megawatts of electricity (1,000 MWe).

Gas-cooled reactor – A nuclear reactor in which a gas is the coolant. In such a reactor graphite is often used as the moderator.

Gaseous diffusion – A method of isotopic separation based on the fact that gas atoms or molecules with different masses will diffuse through a porous barrier (or membrane) at different rates. The method is used to separate ^{235}U from ^{238}U; it requires large gaseous diffusion plants and enormous amounts of electric power.

Graphite – A very pure form of carbon used as a moderator in some nuclear reactors.

Half-life – The time in which half the atoms of a particular radioactive substance disintegrate to another nuclear form. Measured half-lives vary from millionths of a second to billions of years.

Heavy water – Water containing significantly more than the natural proportion (one in 6500) of heavy hydrogen (deuterium) atoms to ordinary hydrogen atoms. Heavy water is used as a moderator in some reactors because it slows down neutrons effectively and also has a low cross-section (probability) for absorption of neutrons.

Heavy-water moderated reactor – A reactor that uses heavy water as its moderator. Heavy water is an excellent moderator and thus permits the use of inexpensive natural (unenriched) uranium as a fuel.

Hydrogen – The lightest element, number one in the atomic series. It has two natural isotopes of atomic weights one and two. The first is ordinary hydrogen, or light hydrogen; the second is deuterium, or heavy hydrogen. A third isotope, tritium, atomic weight three, is a radioactive form produced in reactors.

Hydrogen bomb – A nuclear weapon that derives its energy largely from fusion (thermonuclear bomb).

Isotope – One of two or more atoms with the same atomic number (the same chemical element) but with different atomic weights.

Light water – Ordinary water (H_2O), as distinguished from heavy water (D_2O).

Light-water reactor – A reactor that uses ordinary water as moderator or coolant. There are two types – boiling-water reactor and pressurized-water reactors.

MWe – One million watts of electricity.

Megaton energy – The energy of a nuclear explosion which is equivalent to that of an explosion of one million tons (or 1,000 kilotons) of TNT.

Megawatt-day per ton – A unit used for expressing the burn-up of fuel in a reactor: specifically, the number of megawatt-days of heat output per metric ton of fuel in the reactor.

Moderator – A material, such as ordinary water, heavy water, or graphite used in a reactor to slow down fast neutrons to increase the probability of further fission.

Molecule – A group of atoms held together by chemical forces.

Natural uranium – Uranium as found in nature, containing 0.7% of ^{235}U, 99.3% of ^{238}U, and a trace of ^{234}U.

Neutron – An uncharged elementary particle with a mass slightly greater than that of the proton, and found in the nucleus of every atom heavier than hydrogen.

Neutron capture – The process in which an atomic nucleus absorbs or captures a neutron.

Nuclear device – A nuclear explosive used for peaceful purposes, tests or experiments. The term is used to distinguish these explosives from nuclear weapons, which are packaged units ready for transportation or use by military forces.

Nuclear energy – The energy liberated by a nuclear reaction (fission or fusion) or by radioactive decay.

Nuclear explosive – An explosive based on fission or fusion of atomic nuclei.

Nuclear-power plant – Any device or assembly that converts nuclear energy into useful power. In a nuclear electric power plant, heat produced by a reactor is used to produce steam to drive a turbine that in turn drives an electricity generator.

Nuclear reactor – A device in which a fission chain reaction can be initiated, maintained, and controlled. Its essential component is a core with fissionable fuel. It usually has a moderator, a reflector, shielding, coolant, and control mechanisms.

Nuclear weapons – A collective term for atomic bombs and hydrogen bombs. Any weapon based on a nuclear explosive.

Nucleus – The small positively charged core of an atom. It is only about $1/10\,000$ the diameter of the atom but contains nearly all the atom's mass. All nuclei contain both protons and neutrons, except the nucleus of ordinary hydrogen, which consists of a single proton.

Nuclide – Species of atom characterized by the number of protons and the number of neutrons in its nucleus.

Plutonium – A heavy, radioactive, man-made, metallic element with atomic number 94. Its most important isotope is fissionable ^{239}Pu, produced by neutron irradiation of ^{238}U. It is used for reactor fuel and in weapons.

Power reactor – A reactor designed to produce useful nuclear power, as distinguished from reactors used primarily for research or for producing radiation or fissionable materials.

Pressure vessel – A strong-walled container housing the core of most types of power reactors: it usually also contains a moderator, reflector, thermal shield, and control rods.

Pressurized-water reactor – A power reactor in which heat is transferred from the core to a heat exchanger by water kept under high pressure to achieve high temperature without boiling in the primary system. Steam is generated in a secondary circuit.

Production reactor – A reactor designed primarily for large-scale production of ^{239}Pu by neutron irradiation of ^{238}U. Also a reactor used primarily for the production of radioactive isotopes.

Proton – An elementary particle with a single positive electrical charge and a mass approximately 1 837 times that of the electron. The nucleus of an ordinary or light hydrogen atom. Protons are constituents of all nuclei. The atomic number of an atom is equal to the number of protons in its nucleus.

Radioactive contamination – Deposition of radioactive material in any place.

Radioactive waste – Equipment and materials (from nuclear operations) which are radioactive and for which there is no further use.

Radioactivity – The spontaneous decay or disintegration of an unstable atomic nucleus.

Reactor cooling pond – A mass of water in which reactor fuel elements are placed, immediately after being taken out of the reactor core, to allow the radioactive fission products to decay. The elements are eventually transported from the pond to a reprocessing plant.

Reflector – A layer of material immediately surrounding a reactor core which scatters back or reflects into the core many neutrons that would otherwise escape. The returned neutrons can then cause more fissions and improve the neutron economy of the reactor. Common reflector materials are graphite, beryllium and natural uranium.

Research reactor – A reactor primarily designed to supply neutrons or other ionizing radiation for experimental purposes. It may also be used for training, materials testing and production of radionuclides.

Safety rod – A standby control rod used to shut down a nuclear reactor rapidly in emergencies.

Spent fuel element – Nuclear reactor fuel element that has been irradiated to an extent that it can no longer effectively sustain a chain reaction.

Spontaneous fission – Fission that occurs without an external stimulus.

Thermal-to-electric conversion efficiency – The ratio of the electric power produced by a power plant to the amount of heat produced by the fuel; a measure of the efficiency with which the plant converts thermal to electrical energy.

Thermal (slow) neutron – A neutron in thermal equilibrium with its surrounding medium. Thermal neutrons are those that have been slowed down by a moderator to an average speed of about 2 200 metres per second (at room temperature) from the much higher initial speeds they had when expelled by fission.

Thermonuclear reaction – A reaction in which very high temperatures bring about the fusion of two light nuclei to form the nucleus of a heavier atom, releasing a large amount of energy. In a hydrogen bomb, the high temperature to initiate the thermonuclear reaction is produced by a preliminary fission reaction.

Thorium – A naturally radioactive element with atomic number 90 and, as found in nature, an atomic weight of approximate-

244

ly 232. The fertile ^{232}Th isotope is abundant and can be transmuted to fissionable ^{233}U by neutron irradiation.

TNT equivalent – A measure of the energy released in the detonation of a nuclear explosive expressed in terms of the weight of TNT (the chemical explosive, trinitrotoluene) which would release the same amount of energy when exploded. It is usually expressed in kilotons or megatons.

Transuranic element (isotope) – An element beyond uranium in the Periodic Table, that is, with an atomic number greater than 92.

Uranium – A radioactive element with the atomic number 92 and, as found in natural ores, an average atomic weight of approximately 238. The two principal natural isotopes are ^{235}U (0.7% of natural uranium), which is fissile, and ^{238}U (99.3% of natural uranium) which is fertile. Natural uranium also includes a minute amount of ^{234}U.

Weapons-grade plutonium (uranium) – Plutonium containing at least 90 per cent of ^{239}Pu and no more than 10 per cent of other plutonium isotopes. Uranium enriched to at least 95% ^{235}U.

Yellow cake – Uranium U_3O_8.

Yield – The total energy released in a nuclear explosion. It is usually expressed in equivalent tons of TNT (the quantity of TNT required to produce a corresponding amount of energy).

Source: Based on *Nuclear Terms*, *A Brief Glossary*, (US-Atomic Energy Commission, Division of Technical Information, Oak Ridge, Tennessee, 1967).